Has SCIENCE Displaced the SOUL?

Has SCIENCE Displaced the SOUL?

DEBATING LOVE AND HAPPINESS

KEVIN SHARPE
with REBECCA BRYANT

ROWMAN & LITTLEFIELD PUBLISHERS, INC.
Lanham • Boulder • New York • Toronto • Oxford

ROWMAN & LITTLEFIELD PUBLISHERS, INC.

Published in the United States of America
by Rowman & Littlefield Publishers, Inc.
A wholly owned subsidiary of The Rowman & Littlefield Publishing Group, Inc.
4501 Forbes Boulevard, Suite 200, Lanham, Maryland 20706
www.rowmanlittlefield.com

PO Box 317
Oxford
OX2 9RU, UK

Distributed by National Book Network

Library of Congress Cataloging-in-Publication Data

Sharpe, Kevin J.
Has science displaced the soul? : debating love and happiness / Kevin Sharpe with Rebecca Bryant.
p. cm.
Includes bibliographical references and index.
ISBN 0-7425-4264-5 (cloth : alk. paper)
1. Happiness—Religious aspects. 2. Love—Religious aspects. 3. Psychology and religion. 4. Evolutionary psychology. I. Bryant, Rebecca, 1970– II. Title.
BL65.H36S53 2005
202'.2—dc22 2004023942

Printed in the United States of America.

∞™ The paper used in this publication meets the minimum requirements of American National Standard for Information Sciences—Permanence of Paper for Printed Library Materials, ANSI/NISO Z39.48-1992.

To Mum and Dad, on whose shoulders
I stand and who taught me much about
love and happiness

CONTENTS

PREFACE

Love and happiness represent two emotions lying at the heart of human existence. Ask anyone what they most desire from life, and they will likely reply simply with "happiness." We go to great efforts to satisfy this desire: seeking out engaging and fulfilling occupations, choosing and decorating our homes for maximum comfort and contentment, filling our leisure time with enjoyable pursuits, and making lifestyle choices that we hope will make us happy. Much of our energy goes toward finding a suitable mate, someone with whom we can share our hopes, our fears, our desires, someone with whom we can attain utmost happiness. Once we have found our heart's desire, many of us decide to strengthen and fulfill that union further by together creating and caring for children. Close loving relationships shared with our partners and, perhaps later, with our children represent one sure route to happiness for many of us.

The centrality or importance of an idea often becomes apparent from the effort we expend in trying to explain or account for it. So it is with love and happiness. These two emotions traditionally fall within the remit of spiritual explanation, but increasingly they are becoming the focus of scientific scrutiny. With the recent growth of behavioral genetics, neurochemistry, and evolutionary psychology, a new window opens onto our behavioral, emotional, and social traits. Love and happiness capture the attention of the two most influential explanatory systems available to humanity: scientific and spiritual.

My aim in this book is threefold. I first examine spiritual and scientific approaches to love and happiness. I show how, in many spiritual traditions, these emotions involve a moral or spiritual relationship between ourselves, our actions, and the Divine (or God). The Divine loves us infinitely, desiring above all else our happiness. The Divine urges us to love our fellow humans, to seek out happiness on earth. Some traditions assure us that ultimate happiness awaits the good and faithful in the life to come. The scientific perspective, by contrast, concentrates on the physical mechanisms underlying love and happiness, ignoring both the Divine and the afterlife. I show how, according to the scientific story, love and happiness derive from our genetic inheritance working in conjunction with naturally occurring chemicals in our bodies and brains. I show how love and happiness represent biological adaptations that encourage us to reproduce, to nurture our young, to maximize our achievements, and to push our potential.

A conflict emerges. I show how spiritual thinkers mount various (largely unsuccessful) campaigns to armor their own discipline and to cast doubt on scientific findings, while scientists remain unrattled, assured of success and public support for their subject. Tantalizing questions arise. How can the Divine, something spiritual, experience biologically rooted emotions? Do love and happiness reduce to nothing but genes, hormones, and neurotransmitters? I contextualize the dispute, explaining how these questions reflect an earlier debate between evolutionary psychology and spiritual thought, a debate concerning the reduction of such things as altruism and morality to merely aids for the passage and survival of our genes. My second aim thus becomes one of highlighting the apparent incompatibility between spiritual and scientific accounts of love and happiness, of cataloging the subsequent dispute between spiritual and scientific thinkers.

A handful of spiritual thinkers attempt to move beyond this impasse. They eschew the dualism employed by those who desire the separation of science and the spiritual, by those who talk of different worlds, different subject matters, different languages, or different levels of explanation. I examine two such attempts—those of Philip Hefner and Ted Peters—in detail. These accounts still leave us, however, with serious unanswered questions. How can the Divine direct a random process like evolution? How can we reconcile the big bang with divine creation out of nothing? Does it make sense to claim that the nonbiological Divine shares in human purposes and desires?

Finally, I offer my own solution to the problems thrown up by the debate over love and happiness. My resolution is radical. Important explanatory roles can and should exist for both science and spiritual thought, but to achieve full integration, we must demolish and reconstruct some of our most trusted conceptions. I offer a novel understanding of the Divine, a naturalistic understanding that equates the Divine with the universe-as-a-whole. This understanding divests the Divine of human characteristics yet paints a picture in harmony with our modern, scientific interpretation of the universe, a picture that may legitimize spiritual thought in the eyes of science. I also propose that scientists and spiritual thinkers adopt an overarching empirical framework, a framework that enables scientific evaluation of spiritual claims about human love and happiness. Finally, by examining the ways in which different wholes relate to their parts, I offer a radical reinterpretation of love and happiness in the divine context.

ACKNOWLEDGMENTS

I wish to acknowledge the immense assistance of Rebecca Bryant in the writing of this book. In a period of my life when I had little time and energy for a project like this, Rebecca helped forge my ideas skillfully and quickly into a coherent manuscript that said what I intended.

I also acknowledge the support of the principal, Ralph Waller, and the faculty, staff, and students of Harris Manchester College, Oxford University, who have given me an intellectual home and openly provided space and resources for me to pursue my research and writing.

My family—my daughters, Miriam and Kiri; parents, Pam and Jim; sister, Karen; nephew, Nix; and most especially my wife, Leslie—lovingly support me and bring me much happiness.

—Kevin Sharpe

CHAPTER 1

Happiness and Spiritual Traditions

> O happy day, O happy day,
> When Jesus washed our sins away.

Those who find salvation are some of the happiest people around. At least so a band of popular belief suggests. Research supports it.

Many spiritual traditions would agree as well. The idea of happiness abounds in spiritual and philosophical thought, from the teachings of the ancient Greeks through to contemporary Christianity, as well as in such traditions of the East as Buddhism and Taoism. We experience happiness in this life as a result of virtuous thoughts and deeds, and we secure ultimate happiness in the life to come through a combination of that righteousness with faith in God.

SOCRATES

The ancient Greek philosopher Socrates (469–399 B.C.E.) avowed that the soul, or conscience, constitutes the person and that our happiness depends directly on the good or bad state of our soul. We must know what comprises the true good (that is, true happiness) if we want to live happily, says Socrates, otherwise we will mistake things which aren't really good for the true good. We always act on what's truly good, he says, once we know what this is; it makes no sense to do otherwise and favor misery over happiness. Theodore Kaczynski, better known as the Unabomber, remained true to his anarchist ideals, publicizing his views via a string of bombings that he plotted from his handmade cabin in Montana and, even more amazingly, eluding capture for seventeen years. He might consider himself happy; no way, says Socrates. Kaczynski can't experience true happiness because his *conscience* needs attention. He believes in certain ideals and became powerful though acting violently on those ideals, but misguided beliefs, power, and violence don't equate with the true good or true happiness. The Unabomber mistakes these things for happiness and heads down the rocky road to misery.

PLATO

Other early philosophers built on Socrates' notions of objective goodness and knowledge of the good leading to enactment. Plato's (427–347 B.C.E.) *Republic* introduces a discussion of the nature of justice with the following Tolkienesque puzzle: if a ring renders its wearer invisible, what reasons could the wearer have for acting justly? Invisibility provides the wearer with the chance to get up to all sorts of mischief. Many if not most of us, if presented with such an opportunity, would probably take advantage of it. Wouldn't we nip into Sears for a spot of cost-free shopping, for example?

Platonic justice involves a perfect balance between the three elements of the soul: reason, emotion, and desire. In a just (and so good) person, the soul exists in perfect harmony, with reliable reason governing over the more unpredictable sides of emotion and desire. The good person finds great joy through pursuit of knowledge. In fact, according to Plato, the greatest happiness in this life comes from intellectual speculation or "doing" philosophy. Philosophy professors can appear some of the most rational, emotionally barren people on the block; many students might agree. Would they also agree that this state can lead to happiness?

Wearers of the magic ring would want to act justly, concludes Plato, for the sake of justice alone. They would want to do right because it's right, because in this way we would feel at peace with ourselves and with the gods. Intellectual speculation tells us so. Goodness or justice in this life leads to happiness.

These thoughts extend to the way Plato views the afterlife. The human soul is immortal, and the just will receive their greatest rewards in the life to come, even if poverty, discomfort, or illness mar their current lives. Plato's *Phaedo* indicates that the philosopher genuinely experiences the final goal—purity of wisdom—only on fully quitting the body.[1] Bodily desires and frustrations act like a straitjacket and restrict our capacity for happiness. The true philosopher attains utmost joy only on retracting from the senses and from carnal distractions.

ARISTOTLE

In his *Ethics*, the later Greek philosopher Aristotle (384–322 B.C.E.) moves away from impersonal moral absolutes (doing right because it's right) to what is most conducive to a person's good. He, like Socrates, nominates gaining happiness as the highest good and characterizes it as the soul acting in accordance with virtue. Children enter our world with the capacity to learn both moral virtues (that the irrational elements of the soul—emotions and desires—govern) and intellectual virtues (that the rational elements of the soul govern). Of these, like Plato, Aristotle proclaims intellectual contemplation the greatest happiness available to us. Reason constitutes the highest virtue humans possess since it distinguishes us from other animals. Further, happiness is the soul acting in accordance with virtue, and so the greatest happiness must accord with the highest virtue. A thirst for knowledge and love of books leads to greater happiness—because it expresses a higher virtue—than vegetating in front of our favorite game show or demolishing another TV dinner. Aristotle goes further than Plato, though: acting in accordance with moral virtues can also produce happiness but of lesser quality. Thirst for knowledge even outstrips doing a good turn for a stranger.

The kind of themes raised by the ancient philosophers enjoys a new lease of life in the work of early Christian thinkers.

AUGUSTINE

Augustine of Hippo (354–430 B.C.E.) supplements the Christian Scriptures and their moral commands with a cogent, methodical philosophy of ethics. He transforms Plato's view of the intellectual soul into a Christian view in which we identify with our souls while using our bodies to obtain spiritual ends. All the good deeds we uncomplainingly perform in this life—giving money to charity or helping a friend in need—add up to a one-way ticket to our future happiness. Happiness persists as the greatest human goal, claims Augustine, but this takes on a new meaning: reconciliation of the soul with God in the life to come. Augustine thus shifts the emphasis to happiness in the hereafter. We struggle in this life to gain eternal happiness in the next. Down with academia, he chants. Philosophical or intellectual contemplation can never lead to true wisdom and happiness. Up with altruism, he cries. Only loving God and one's neighbor (living a morally virtuous life) guarantees a future of never-ending happiness by God's side.

AQUINAS

The Scholastic philosopher Thomas Aquinas (1225–1274) also seeks to reconcile the philosophy of the ancient Greeks with Christian doctrine, and his work became part of the ideology of the Roman Catholic Church. Like Aristotle, Aquinas adopts a teleological view of morality (in which we direct all our efforts toward one ultimate moral goal), naming happiness as the ultimate end we seek to attain. Like Augustine, Aquinas replaces "intellectual contemplation" with "love of God" in the equation "greatest happiness = intellectual contemplation." We strive to know God in this life, but our struggle only ceases in the life to come. Heaven becomes our reward for loving God. It's the only source of true happiness.

Aquinas does diffuse this position a little: by living a virtuous life, we attain some degree of happiness, albeit inferior to the purity of eternal happiness. We gladly accept refreshments on route to the winning post, like athletes running the New York marathon, but, once we cross that line, the euphoria of success eclipses everything else.

The teachings of Augustine and Aquinas color contemporary Christianity. Their theorizing about happiness permeates today's spirituality.

CHRISTIANITY

Contemporary orthodox Christianity loses sight of the here and now, focusing instead on happiness lying someplace else—a land of original bliss and innocence (the Garden of Eden) or of future joy (Heaven, our eternal and happy home where we will see God face-to-face or the Promised Land where we will find happiness and complete satisfaction). "Heaven is destination and reward," writes David van Biema, "succor and relief from earthly trials."[2] Adds Jeffrey Russell from the University of California at Santa Barbara, "[Heaven] is an endless dynamic of joy."[3] A friend with a staunch Roman Catholic upbringing talks of her constant sinning because she fails to say grace before every meal, pray every night, and attend church as often as possible. She feels she must overcome this tendency through acts of penance to achieve happiness in the afterlife.

Modern spiritual leaders like Robert Schuller prefer to focus on happiness in the present. He writes about *The Be Happy Attitudes: Eight Positive Attitudes That Can Transform Your Life*. Such charismatic and Pentecostal movements assume that the spiritual intends for happiness. Happiness is nearness to God. We move close to God through the "happy clappy" world of human togetherness, epitomized by hallelujah crying and hymn singing.

And today's believers do stand out as prime examples of happy people. The highly spiritual declare themselves very happy at twice the rate of those with the lowest spiritual commitment, according to a recent Gallup survey. A study of 166,600 people in fourteen countries demonstrates that happiness and satisfaction with life increase with frequency of attendance at worship services.[4]

Modern spiritual thinkers propose that an active and committed spiritual life leads to happiness. To their eyes and to those of millions of contemporary Christians, the Bible paints a picture of a gracious and loving deity who desires everyone's happiness. Happiness arises directly from God.

Not only do Western spiritual traditions and their philosophies focus on happiness. This motif plays a significant role in Eastern spiritual thought too.

CONFUCIANISM

The ancient Chinese philosophy of Confucianism (Confucius, 551–479 B.C.E.) teaches that not pleasure, honor, or wealth but virtue alone produces true happiness. Much weight rests on the notion of the family, where the life of each individual continues those of his or her ancestors, a link in the familial chain. Pursuit of happiness in the next life isn't the primary concern. Confucianism instead encourages us to seek to expand, revitalize, and strengthen our families by living frugally, working hard, spurning selfish desires, maintaining a respectable position in society, and generally leading a virtuous life. Self-control, self-improvement, and the practice of moral virtues dominate.

Welfare of the individual therefore gives way to welfare of the family, social group, and even the entire human race for the Confucian. Individuals must share their successes with the group. Happiness results from "knowledge, benevolence, and harmony of the group."[5] The suburban yuppie, cell phone in hand, enclosed in an SUV, and dominated by the underlying slogans of "greed is good" and "all for me" represents the antithesis of Confucianism, the Confucian's worst nightmare. We can achieve ultimate happiness by giving to society and thus by living virtuously.

TAOISM

Taoism represents a second major school of thought in China that influenced art, literature, and spirituality there. Taoists teach that ultimate happiness arises from gaining the original perfection of the Tao (or the Way), encompassing the ideal way of life in which we live in perfect harmony with the forces of nature. Early Taoists believed that their ancient ancestors possessed the primitive Way but later lost it. The Tao now needs rediscovering. The Way lauds qualities such as receptivity, passivity, and humility, and it advises a life of natural simplicity and detachment from worldly pursuits. What we call happiness isn't true happiness; money, power, beauty, fast cars—the things that we ordinarily pursue with a vengeance—won't bring it. Ultimate happiness for the Taoist consists of a down-shifting from this rat race of materialist living. Ultimate happiness consists of "inaction." One of the earliest Taoist texts, the *Chuang-tzu* (Master Chuang), goes so far to suggest that perfect happiness can't be found on earth.

Happiness occupies the pole position in Chinese life, and they venerate Fu Shen, the god of happiness (the bat [*fu*] symbolizes happiness). They believe that human beings can't obtain happiness solely through their own actions because *Ming* (or fate) plays a role.

BUDDHISM

The Buddha (563–483 B.C.E.) preached four "noble truths":

- Life is suffering.
- Suffering involves a chain of causes.
- Suffering can cease.
- A path to cessation of suffering exists.

We all suffer because we want to retain a sense of our own identity, yet everything in the world is impermanent or transitory and lacks a soul. People comprise aggregates or bundles of psychological elements (thoughts, feelings, sensations) and physical elements (arms, legs, head, torso); no "person" or self exists over and above the constituent parts, and nothing acts as a container for these parts. Our longing for personal identity leads to a repeated cycle of death and rebirth (the "Wheel of Becoming") that causes pain and suffering.

Understanding that we don't have a permanent self or ego leads to cessation of suffering, according to Buddhist thought. By abandoning all forms of worldly desire (including the desire to attain release from the Wheel of Becoming), we move toward *Nirvana,* the final release from the cycle of rebirth. The path to *Nirvana* requires us to adopt the virtues of clarity, lack of desire, and universal compassion and friendliness. We finally reach *Nirvana* through selflessness, literally through annihilation of the illusion of the self. *Nirvana* represents freedom from desire and frustration, pain, and suffering. It represents ultimate happiness.

Happiness features in Buddhism in two other ways. First, the word *sukha* denotes pleasant bodily and mental feelings, important because it facilitates successful meditation. A happy mind is a concentrated mind, for the Buddhist. *Sukha* stands in opposition to *dukkha,* which conveys the suffering and pain of mortal existence.

Buddha *Amitabha* rules over *Sukhavati* ("the Blissful"), his creation. People who worship *Amitabha* and repeat his name may be reborn in *Sukhavati* and live there blissfully before entering *Nirvana*. This introduces yet another way that happiness features in Buddhism. The *sutras* (practical summaries of the Vedic scriptures) devoted to *Amitabha* describe *Sukhavati* as a paradise filled with glory from *Amitabha*: exotic, bounteous, filled with beautiful flowers, trees of jewels, and wonderful fragrances. Rivers of sweet waters flow by carrying flowers. Their flow is music. Beauty and comfort reign. No corpses, beasts, or hells mar the sumptuous landscape. All our wishes are granted here, and we no longer experience pain or sadness. Waking in a lotus flower represents rebirth in this *Sukhavati* for the faithful. The greatest happiness arises from hearing *Amitabha* preaching before our final entry into *Nirvana*.

We should understand *Sukhavati,* in its deepest sense, as depicting a state of mind. The descriptions, though graphic, are only figurative. *Sukhavati* seems to share much in common with the Christian conception of Paradise, and like Paradise it doesn't represent the ultimate destination or perfect happiness, merely a halfway house. Only the nothingness of *Nirvana* equals escape from the suffering of life. Only *Nirvana* equals pure happiness.

HINDUISM

Hinduism also advocates withdrawal from the world of pleasure or desire, but its underlying philosophy differs from that of Buddhism.

Hindus believe in universal determinism or the doctrine that past events determine future ones (Saddam Hussein's invasion of Kuwait resulted several years later in the U.S. and British military invasion of Iraq, for example). They also believe in the reincarnation or rebirth of individual souls to eternal life. These beliefs merge in the doctrine of karma: we bring all suffering and happiness on ourselves: "what goes around, comes around." Performing a morally good deed in this life notches up greater happiness for the next and vice versa for an evil action, according to the great celestial abacus. Hindu philosophy places literal interpretations on karma; slanderers are supposedly reincarnated with bad breath, for example. Karma ensures that it is not God but us who predetermine the quality of our own future lives.

An error on our part leads to our suffering. The Samkhya system of Hindu thought distinguishes the world's many selves from the one true world: many eternal centers of pure consciousness versus a state of continuous flux. Nature, the true world, comprises illumination, kinesis, and inertia, and we experience these three elements as pleasure, pain, and torpor. Our centers of consciousness mistake themselves for more tangible entities like human bodies, sense organs, and intellects. They compound this error by considering the tangible entities to experience pleasure, pain, and torpor. But these entities actually belong to the unity of nature, and our experiences rest in pure consciousness alone. This mistake leads to our suffering.

We can arrest our suffering, suggests the Yoga system of Hindu thought. To do this, we must liberate our minds from their objectifying tendencies and discriminate between the self and nature in meditation. This allows the self to return to pure consciousness. Our suffering ceases, and we attain happiness.

ISLAM

Islam also considers happiness of great importance:

> By the soul and [him who] formed it,
> And implanted into it its wickedness and its piety!
> Blessed is he who purifies it.
> Ruined is he who corrupts it.[6]

This passage taken from the Qur'an (the holy book of Islam) demonstrates the Islamic belief that we are born, as clean slates, with the capacity for both good and evil. God molds this capacity by testing us throughout our lives: some of us opt for good, others for evil. God acknowledges our choices in the form of eternal reward or punishment at the Last Judgment: "And we try you with evil and good as a test; then unto us you will be returned."[7] Those receiving eternal reward enter *Al-janna* (the garden), the home of the blessed, where they live in luxury surrounded by rivers of milk, purified honey, and sweet-tasting wine. They relax on couches, attended by wide-eyed maidens, and sport fine, richly woven garments. Life there lasts forever.

Islam, unlike either Hinduism or Buddhism, understands *Al-janna* (the afterlife) as a paradise where the righteous enjoy the highest of spiritual and sensual happiness. Muhammad's (570–632) prophecy captures the spirit: in Paradise believers will own a pearl tent, sixty miles long, with a family in each corner, and two gardens containing silver and gold vessels.

JUDAISM

The spirit of God, according to a Talmudic saying, doesn't rest on a gloomy person. In the world to come, according to another Talmudic rabbi, we will need to account for whether we used appreciatively each enjoyment this life offered us or whether we declined them gratefully. Jewish legend portrays the prophet Elijah telling Rabbi Beroka, a proudly austere man, that he would have two jesters as companions in Paradise because they would cheer the depressed and bring more joy into the world. We ought to enjoy the Sabbath and other holy days, and we should adopt whatever legitimate methods (like good food and good clothes) would make them so because gladness brings the spirit of God near to us.[8]

Abstinence is appropriate in our lives, according to the teachings of Judaism, when it curbs wild passions and desires. Happiness, as with all enjoyments and pleasures, has a rightful place provided we place it second to service to God. This means that we must subject it to moral and spiritual considerations, such as moderation, and sacrifice it when a higher principle requires this.

CONCLUSION

The major spiritual traditions invoke ideas of happiness. Many speak of happiness in the afterlife, while others discuss happiness in the here and now. Some involve notions that people in other traditions may find alien, even incomprehensible. How can a Christian understand the Buddhist idea that the void of *Nirvana* represents ultimate happiness? Despite these differences, happiness plays a pivotal role in spiritual traditions, and they describe its nature in spiritual terms.

How else might we describe happiness? Which fields, besides spiritual thought, might provide further insights into the nature of happiness? The next chapter will examine some radically different approaches to describing and uncovering the nature of human happiness.

CHAPTER 2

Happiness and Contemporary Science

> She thought now, as with eyes closed she floated, oh how perfect this is, oh I am so happy! And yet, some other nearby thought-self was saying, how can I be happy now, when everything is going very soon to be dissolved into pieces and made as if it had never been.[1]

We all desire happiness and devote considerable time to seeking it. How to secure a job that will satisfy and fulfill us? How to find a partner with whom we can happily share the rest of our lives? How to ensure that we spend our leisure time pleasurably and constructively?

Spiritual belief or faith represents one way to attain the happiness prize, as the previous chapter discussed. Yet whether Jesus or anyone else saves us or not, most of us already feel happy and satisfied with life. Surveys of 1.1 million people from all over the globe say, according to psychologists David Myers of Hope College in Michigan and Ed Diener of the University of Illinois, that 93 percent of people feel happy (which includes very happy, pretty happy, and moderately happy) as opposed to sad or neutral. Most of us describe ourselves as "pretty happy."[2] Holding to a particular or any spiritual belief doesn't seem to matter, happiness-wise.

If spiritual thinking doesn't have the answer, on what does happiness depend?

THE HAPPINESS "SET RANGE"

For each of us, our happiness fluctuates within a small range called a "set point" or "set range" that our genes largely determine. So concludes molecular biologist Dean Hamer of the National Cancer Institute in his review of studies on the role of genes in happiness or misery.[3] The set range represents a kind of preset value with which we are born and to which our level of happiness inevitably returns. This notion resembles the metabolic set range that some scientists claim governs our weight; no matter how many cakes or chocolates we eat, the body's metabolism readjusts to maintain its preset weight. This could explain why some people find it so hard to shed excess pounds, while others are lucky enough to have figures like supermodels Claudia Schiffer and Kate Moss. Studies showing that body mass is 70 percent heritable lend credence to the metabolic set range.[4] Some of us similarly always approach life full of hope and enthusiasm, while others seem permanently to experience the blues. Though we experience temporary mood swings, we soon readjust to our genetic set range for happiness in the same way as with our weight fluctuations.

Support for the genetic set range comes from a series of twin studies conducted by behavioral geneticist David Lykken and psychologist Auke Tellegen, both at the University of Minnesota. Twins provide an excellent base from which to study the degree of heritability of behavioral traits because identical twins share identical genes, whereas fraternal twins share genes as do ordinary siblings (roughly 50 percent). Take a particular characteristic and find out how often identical twins share it and how often fraternal twins do. The greater the difference between these two percentages, the greater the characteristic's heritability and the smaller the role played by external or environmental factors.

Identical twins attain the same level of happiness 44 percent of the time, while fraternal twins reach the same level only 8 percent of the time, according to Lykken and Tellegen's research. Their study asked 1,380 pairs of twins raised together to rate themselves on claims like "I am just naturally cheerful" and "My future looks very bright to me." Similar results emerge from a smaller previous study with twins separated in infancy and raised separately. "This conclusion means that the variance in adult happiness is determined about equally by genetic factors and by the effects of experiences unique to each individual," say Lykken and Tellegen. Sex, age, race, and marital status of the twins have only a slight (2 percent) impact. Hamer echoes these claims: "These data show that the broad heritability of well-being is 40 to 50 percent." Heritability raises even higher for happiness in the long term. Lykken and Tellegen administered the same questionnaire to a subset of the twins five to ten years later and then performed cross-twin, cross-time calculations, comparing the score of one twin at 20 with his or her cotwin at 25 or 30. The correlation statistics show, Lykken and Tellegen write, that the "heritability of the stable component of well-being is about .80."[5] A twin's self-report of well-being provides a good indicator of the state of the other twin. Thus, according to Hamer, "How you feel right now is about equally genetic and circumstantial, but how you will feel on average over the next ten years is fully 80 percent because of your genes."[6]

With heritability this high, wealth, education, or social status say surprisingly little about a person's happiness.

Different types of research similarly show that a person's level of happiness remains stable over many years. In a study of 5,000 adults carried out by the National Institute on Aging in America, people who felt happiest in 1973 showed up as relatively happy ten years later. Changes that we naturally associate with major emotional upheaval—like starting a new job, getting married, or moving house—make no difference to happiness levels; scores for people who had experienced these changes remained as stable as for people whose situation stayed much the same.[7] Psychologists Christopher Lewis at the University of Ulster and Stephen Joseph at the University of Essex more recently report that the Depression-Happiness Scale (which psychologists use to calculate levels of happiness) measures happiness as an underlying trait rather than, which researchers used to assume, as a transitory state. Lewis and Joseph's work shows that scores on the scale remain relatively stable over two years. This in turn suggests that people's happiness levels persist over the long term.[8]

HAPPINESS AND THE UPS AND DOWNS OF LIFE

Maybe our happiness does stick at roughly the same level in an average everyday life. But what about the momentous events—winning the lottery or giving birth to a first child—and the desperately tragic—losing a spouse or becoming permanently paralyzed? Do these have a long-lasting effect?

Consider Rose Marie Lajoie, a Michigan lottery winner. She says, "If you are a negative person to start off, if you are a dull person to start off, you'll be the same way [after winning the lottery]."[9] Momentous events alter our level of happiness for a short time—the 50 percent nongenetic variation in happiness over the short term allows for this—but we quickly adapt, so the long-term set range remains unaltered. We all recognize that euphoric feeling when we attain something precious—the coveted job or the college degree, perhaps—yet the feeling doesn't stay with us long. All too soon we forget and move on, our eyes firmly fixed on the next target. "So many people plan their lives for a distant goal," Lykken explains. "They believe that if they become C.E.O. or win a gold medal, their lives will rise out of humdrum ordinariness. This isn't so. There's a rush of glory and then it fades."[10]

The sting of tragedy disperses equally as fast. Research by Diener and fellow University of Illinois psychologist Carol Diener indicates that even quadriplegics and others with severe disabilities describe themselves as happy. In their more objective reports, they can remember more good than bad events in their lives and say they experience more positive than negative emotions day to day. Reports from friends, family, and interviewer ratings corroborate these findings.[11] A study of car accident victims in Michigan reports that, only three weeks after suffering a paralyzing spinal cord injury, victims feel happiness as the overriding emotion.[12] Another study by researchers at Northwestern University and the University of Michigan compares a sample of Illinois lottery winners, individuals who had suffered crippling accidents, and a control group that had escaped both fates. The lottery winners generally feel less happy than the control group, and the disabled people feel much more happy than expected.[13] All this evidence supports Lykken's prediction that Christopher Reeve (the late movie star paralyzed by falling from a horse) was probably as happy after his accident as before.[14]

Diener puts a time on how long it takes people to adapt to relatively minor events like gaining a promotion or losing a lover. "The effect on people's mood is gone by three months, and there's not a trace by six months," he says. Expect the effect to have dispersed within a year. For more serious events like divorce, bereavement, or unemployment, the effects can last longer, of course. This tends to indicate a clinical disease, such as depression, which overrides the customary set range. "It's because in some sense," says Diener, "the bad event continues to happen—there are reminders every day."[15]

"The 'slings and arrows of outrageous fortune' clearly influence mood," explains Greg Carey, a behavioral geneticist at the University of Colorado, "but long-term equilibration to life's ups and downs is partly a function of the slings and arrows of genetic fortune."[16]

THE BIOCHEMISTRY AND NEUROLOGY OF WELL-BEING

The genetic view of happiness has implications for our understanding of the cause of our feelings of well-being because our genetic code translates directly into how our neurology (nervous system) behaves.

Hamer directs our attention to two of the more than three hundred known neurotransmitters: dopamine and serotonin. Dopamine acts as the brain's chemical for pleasure. "It is what is released after a good meal, a pleasant sexual experience or a hit of cocaine," Hamer explains.[17] Recreational drugs like amphetamines prove so popular because they belong to the same family as dopamine and produce similar effects: feelings of happiness, contentment, and satisfaction.

Hamer describes serotonin as "the brain's punishment chemical"; with its reduced activity, misery appears. Scientists associate lack of serotonin with depression, suicide, and anxiety. These are symptoms of a modern malaise. As Elisabeth Wurtzel describes in her book *Prozac Nation: Young and Depressed in America,* doctors now prescribe the drug Prozac (which prolongs the action of serotonin produced by the brain) as a matter of course to counter these negative emotions.

Neurotransmitters like dopamine and serotonin work by passing information from the synapse or junction between a nerve cell and another nerve cell or a muscle. The nerve cell's bulbous end releases them from storage when an electrical impulse moving along the nerve reaches it. They then cross the junction to dock at a receptor on the other nerve cell, like spacecraft docking at a space station, and either prompt or inhibit the impulses along the second cell. The first nerve cell reabsorbs excess neurotransmitters but not necessarily all of them. Those that remain free floating, according to biology, help create our happy or miserable states of being.

Genes carry the instructions for the construction of neurotransmitters and their receptor and reabsorption portals. They also impart information on such things as their storage and release rates. Hence, genes can influence the prevalence, scarcity, and activity of serotonin and dopamine and, in turn, whatever behaviors and feelings these neurotransmitters induce. Researchers have found, for instance, that people who differ in the gene that produces part of the D4 dopamine receptor—the part that controls the amount of dopamine binding there—differ in a parallel way in their moods. People with the highest levels of dopamine report feeling the most positive. Psychologist Richard Davidson at the University of Wisconsin comments that this is "the first time there's been a specific connection between a molecular genetic finding and people's levels of happiness."[18]

Some scientists think they have located the part of the brain that registers happiness and where the set-range mechanism works. Davidson has found that people with more activity on the left prefrontal area of the brain experience greater happiness, while those with greater activity on the right prefrontal area experience more negative emotions. People with the greatest right prefrontal activity suffer from clinical depression and claim that life holds no pleasure for them. Even very young children appear to fit the pattern: babies of ten months tend to cry less easily when separated from their mother for short periods if they exhibit more active left prefrontal lobes. Further evidence derives from the work of Richard Lane and his colleagues at the University of

Arizona's College of Medicine. Their preliminary research indicates that feelings of happiness, sadness, and disgust all co-occur with increased brain activity in the thalamus and medial prefrontal cortex. Greater activity near the ventral medial frontal cortex distinguishes happiness from sadness, while happiness correlates with significant increases in bilateral activity near the middle and posterior temporal cortex and hypothalamus. Lane concludes that "spatially distributed brain regions participate in each emotion."[19]

ATTITUDES FOR HAPPINESS

Other sciences beyond behavioral genetics and neuroscience contribute to this discussion. Social psychology explores activities that activate our happiness: sharing in stellar sex or consuming delicious dinners, perhaps. Myers lists four character traits that make for happiness:[20]

1. Happy people have high self-esteem; they like themselves. Eighty-five percent of U.S. residents voted "having a good self-image or self-respect" as very important, and 0 percent voted it unimportant, according to a Gallup poll.[21] These kinds of feelings help cushion us against the demons of anxiety and depression and so bolster our happiness levels.
2. Happy people feel optimistic; they exude hope and feel able to succeed at tasks they undertake. Increased optimism means better health, which in turn leads to greater happiness. A study of Harvard University graduates shows that those people who felt the most pessimistic in 1946 were the least healthy in 1980.
3. Happy people are extroverts; they feel self-confident and mix easily with others. Extroverts are more likely to marry, find good jobs, and make close friends, according to research by Ed Diener and Keith Magnus

of the University of Illinois. These achievements lead to greater satisfaction with life.

4. Happy people feel in control of their lives. Allowing prisoners, nursing home patients, and employees to make decisions about their environment and its running results in increases in happiness. Controlling our own time also leads to happiness. Psychologist Michael Argyle of Oxford University comments that happy people "are punctual and efficient," while unhappy people "postpone things and are inefficient."[22] Good time management provides a sense of control.

The "happy farms" scattered across the United States provide commercial counterparts to psychological descriptions of what leads to happiness. Here you can learn about "inner wisdom, self-confidence, personal empowerment, motivation, reconciliation with the past, and greater vitality,"[23] all for a substantial weekly sum. Our determination to find true happiness has turned it into a multi-million-dollar industry.

Mihaly Csikszentmihalyi, professor of psychology and education at the University of Chicago, discusses another road to happiness. In his book *Flow: The Psychology of Optimal Experience,* he writes of when we find ourselves absorbed in an activity and time flies. We then experience flow, he says. Life flows when we engage our skills and talents optimally, avoiding underchallenge (which results in boredom) and overchallenge (which results in stress). When in a state of flow, we feel happy, satisfied, a sense of meaning, purpose, and control.

Csikszentmihalyi first observed this state when studying artists who spent hours absorbed in their work. They concentrated purely on their creation, toiling for the sake of the art alone, not for money, fame, or other extrinsic reward. Numerous other activities besides artistic creation can result in flow, such as climbing a mountain, writing a book, weaving a rug, or playing tennis. Any of us can experience flow, as long as a challenging activity absorbs us. We report more positive feelings when in this state than when we laze around, being bored and doing nothing much. Flow promotes happiness.

Social psychology also tells us about things that fail to make us happy. Happiness doesn't rely significantly on external factors: economic class, age, gender, education, or race. Wealth doesn't correlate with happiness, except in the very poorest countries. Despite the fact that "compared with 1957, [people in the United States] have twice as many cars per person—plus microwave ovens, color TVs, VCRs, air conditioners, answering machines, and $12 billion worth of new brand-name athletic shoes a year"—they feel no happier now than in 1957; 35 percent declared themselves "very happy" in 1957 compared to the slightly smaller figure of 32 percent nearly four decades later.[24] Money doesn't buy us happiness.

HAPPINESS AND EVOLUTION

The science of evolutionary psychology (also known as sociobiology) aims to explain human goals, beliefs, and theories in Darwinian terms—at least in part. The urge to survive and reproduce determines even the ways in which we think, the ways in which our minds work. One leading proponent of evolutionary psychology, Michael Ruse, a philosopher at Florida State University in Tallahassee, words the point bluntly: "Those proto-humans who believed in 2 + 2 = 4, rather than 2 + 2 = 5, survived and reproduced, and those who did not, did not."[25] The belief that 2 + 2 = 4 proves advantageous for our survival; therefore, we take it as true.

Evolutionary psychology has something to say about happiness too. Evolutionary psychologist Donald Campbell describes us as condemned "to live on a hedonic treadmill."[26] We fanatically pursue happiness, yet no sooner do we reach one goal than the satisfaction fades away, and we commence reaching for the next rung on the ladder of pleasure. "As the environment becomes more pleasurable, subjective standards for gauging pleasurableness will rise," he explains, adding that "habituation will produce a decline in the subjective pleasurableness of the input."[27] This, of course, restates the idea of a genetic set range for happiness, as discussed earlier in this chapter. We've seen the scenario before: we feel ecstatic on gaining a pay rise but soon find that our material situation feels little different from before. We no longer feel happy. Perhaps we can live the high life more frequently, but we soon get used to that. We want another rise. We've habituated and feel the need to strive once more.

In an evolutionary scheme, what adaptive advantage did seeking happiness bring to our forebears if frustration and dissatisfaction constitute the net outcome? Campbell suggests the beginnings of an answer: maybe only those people who live in oppression and without hope of motivation have given up entirely on the search for happiness. Other proponents of evolutionary psychology take Campbell's suggestion further: "We are built to be effective animals, not happy ones. . . . Of course, we're designed to *pursue* happiness; and the attainment of Darwinian goals—sex, status, and so on—often brings happiness, at least for a while. Still, the frequent *absence* of happiness is what keeps us pursuing it, and thus makes us productive." So argues Robert Wright in his book *The Moral Animal*.[28]

The *search* for happiness, therefore, plays the key role. From the point of view of evolutionary psychology, our desire for pleasure keeps us on our toes. The activity expands our horizons, our resources, and our skills. Parents employ much the same catch-it-if-you-can psychology when encouraging their offspring to walk; brandishing a favorite toy lures the child into stepping toward it, and moving the toy farther away means that the child progresses a few more steps. As the toy recedes ever farther, the child's walking ability improves proportionally.

A limit blocks how far the pursuit of happiness benefits us, though, just as a limit prevents how far the child can chase the toy before keeling over. As Steven Pinker points out in his book *How the Mind Works*, "The problem is, how much fitness is worth striving for? Ice Age people would have been wasting their time if they had fretted about their lack of camping stoves, penicillin, and hunting rifles, or if they had striven for them instead of better caves and spears."[29] We need to decide what we can reasonably attain. We can gauge this in two ways, according to Pinker: by noticing what *others* have attained and by noticing how well off *we* are at the moment. What others have attained provides an insight into what we might attain for ourselves. This kind of comparison gives rise to the "keeping up with the Joneses" mentality: when Mrs. Smith glances over the fence and sees that Mrs. Jones has a glittering new Humvee, she feels she must have a vehicle just the same or better. We want what others have. The second way that helps us gauge what we can reasonably attain involves our taking stock of how well off we are. We can then aim to achieve just that little bit more, and more, and more. These two standards of comparison help ground evolutionary theory's forecast that our reach should exceed our grasp, "but not by much."[30]

CONCLUSION

The several sciences that discuss human happiness include one common point of focus: happiness is a *natural* phenomenon. This naturalism takes several forms:

- A set range for happiness that our genes encode.
- Neurotransmitters responsible for our states of well-being and misery.

- Concrete activities (engaging in rewarding pastimes or making lasting friendships) leading to joy.
- The pursuit of happiness, bringing adaptive advantages that aid our survival and reproduction.

Despite this apparent diversity of topics, the materialist focus shines through. Our *biology* directs us into our passionate love affair with happiness. No room appears to remain for a spiritual slant on happiness; objectivity replaces subjectivity and mysticism. How, glancing backward, does this affect the issues that chapter 1 examined? Can naturalistic and spiritual accounts of happiness coexist? Should we abandon one account in favor of the other? If so, which one? And why?

Many of us will instinctively feel that happiness must comprise more than biological drives and chemical activity. This sense may feel compelling, and we should take our convictions seriously.

INTERLUDE I

An Unhappy Conflagration

> They can't give us a pill to make us happy. . . . We create our own joys, and we feel happiest in learning to trust each other.[1]

We have met in the past two chapters two divergent accounts of happiness. On the one hand, scientific stories talk materialistically of genes, neurotransmitters, and electrical impulses. On the other hand, spiritual stories talk of intellectual contemplation, virtuous living, and a bounteous afterlife. A clash looms between the two approaches.

DIFFERENCES

Scientific and spiritual approaches to happiness clash in terms of *definitions*:

- Science's understanding of happiness centers on the notion of physical well-being. Scientists measure our happiness levels on the basis of our subjective self-reports of well-being: if we report feeling good, we're happy. We might feel great when consuming our favorite dish or when absorbed in a challenging game of chess, and these feelings constitute our current levels of happiness, according to science.

- Spiritual definitions of happiness, in comparison, concentrate on intellectual or spiritual satisfaction. Happiness, for the spiritual thinker, becomes a loftier affair. We have to work hard for our happiness, whether this involves an intellectual search for the good, a struggle to fill our lives with virtuous thoughts and deeds, or a quest to lose our personal identity on the way to *Nirvana*. Happiness, from the spiritual perspective, represents a noble prize.

Scientific and spiritual approaches to happiness also clash over matters of *timing*:

- Scientists concentrate on happiness in the here and now; they show no interest in happiness in the life to come because such matters lie outside science's remit and the current precepts of its method. Science deals with the material world, so scientists study human happiness only as a physical phenomenon. Once a person's body ceases to function, any happiness the person experiences can no longer be physical, and science must become silent about it.
- Spiritual thinkers also discuss happiness in the here and now but, in comparison, many spiritual traditions—especially those that concentrate on the afterlife—think that nonphysical happiness in the life to come steals the show. After the death of our bodies, we attain ultimate happiness, a seat by God's right hand, our place in the pastures of *Al-janna*, where happiness far surpasses anything we can experience on earth.

How do scientists and spiritual thinkers react to this clash over the nature of human happiness?

REACTIONS FROM SCIENCE

Scientists take little notice of insights from spiritual traditions. Their ignorance reflects, at least partially, the previously mentioned demands of scientific methodology. They consider their job the discovery of facts about the physical not the spiritual world, and so, they conclude, spiritual insights have no bearing on science. The two worlds may coexist, but they're doomed, by definition, never to collide.

The public also provides science with an easy ride: British people trust the word of scientists 59 percent of the time and that of religious organizations only 22 percent, according to a recent study on judging risk.[2] Advertisers take advantage of this "white coat syndrome" embedded deep in our collective psyche. The attitude abounds. Actors parade as scientists proclaiming the efficacy of a product and so convince us to rush to stores to increase the sales of numerous washing powders, toothpastes, and other household essentials and inessentials. When people profess such confidence in your work and abilities, you feel you must be doing the right thing. Complacency easily sets in. Scientists may well feel they have no need to cast a glance at spiritual insights into happiness; they're already heading in the right direction.

REACTIONS FROM SPIRITUAL TRADITIONS

The stakes for a spiritual approach in this conflict rate much higher, and its proponents show far more enthusiasm for defending their territory. Recent work on the science of happiness provokes a number of different reactions from spiritually inclined thinkers.

I've called the first the "Let's-Get-'Em" response. This involves a direct and sophisticated attack on the methods and findings of behavioral genetics.

LET'S-GET-'EM

Results Don't Replicate Researchers cite the existence of single genes responsible for attributes as disparate as manic depression, schizophrenia, alcoholism, and novelty seeking. Journalist Sharon Begley points out that several such claims for the genetic roots of various behaviors run into trouble because follow-up studies fail to replicate the research. Later research doesn't confirm the original results. Sometimes, Begley points out, researchers even retract their original claims. For instance, two separate teams of scientists recently claimed a connection between a version of the D4 dopamine receptor gene, or D4DR, and an adventurous, excitable personality that relishes new experiences. "This work . . . provides the first replicated association between a specific genetic locus involved in neurotransmission and a normal personality trait," the scientists contend.[3] Other teams produced conflicting results. Work by psychiatrist Anil K. Malhotra of the National Institute of Mental Health in Bethesda, Maryland, fails to find a connection between the specific version of D4DR and novelty-seeking behavior in a group of 193 Finnish men. This version of D4DR also varies in frequency in different parts of the world, which makes it an unlikely candidate for causing such a widespread personality trait. "It's unacceptably speculative to claim that large, heterogeneous populations would have an average difference in novelty-seeking behavior just because they differ dramatically in the frequencies of [this gene]," asserts Kenneth K. Kidd of Yale University School of Medicine.[4]

While a number of claims for links between specific genes and personality characteristics may not hold up, others do. Begley fails to mention them. For instance, two independent teams of researchers, one at the University of Utah Health Sciences Center in Salt Lake City and the other at St. Mary's Hospital in Manchester, link a specific gene to an aspect of thought: the deletion of the chromosome 7 gene, LIM-kinase 1, disrupts our ability to visualize and mentally manipulate parts of objects. An activity such as putting together a piece of self-assembly furniture depends on this ability.[5]

Begley further muddies the issue by concentrating on and attacking the so-called OGOD model of genetics, meaning "one gene, one disorder": a single gene alone causes a disease, or, when extended to personality traits like happiness, a single gene underlies one personality trait. However, researchers have now moved away from the OGOD strategy. They usually suggest instead that a configuration of genes shapes any given behavioral trait. Robert Plomin, a behavioral geneticist at the Institute of Psychiatry in London, explains, "Although any one of many genes can disrupt behavioral development, the normal range of behavioral variation is orchestrated by a system of many genes, each with small effects."[6] Begley mentions that a claim for the genetic basis of neuroticism (or anxiety) awaits its follow-up. The journal *Science* has since published the follow-up, and this study confirms that anxiety does have multiple genetic roots. "Variance in personality traits, including those related to anxiety, is thought to be generated by a complex interaction of environmental and experiential factors with a number of gene products," psychiatrist Klaus-Peter Lesch and his coworkers claim.[7] Dean Hamer, one of the authors of the study, later expands, "We think there may be ten genes altogether that influence anxiety. But there may be a hundred or a thousand."[8]

We must, of course, view with caution any study that scientists can't replicate. Yet the picture is brighter than Begley paints it. Researchers do replicate studies. More important, failure to replicate can lead to increased knowledge, as with the realization that, though the OGOD model may not apply to behavioral traits, multiple genes acting in concord do affect behavior. Genetic explanations remain basically sound.

Twin-Studies Skepticism Begley refers to a second line of attack that critics often launch, this time against the study of twins. Much of the genetic-basis-of-happiness research draws on twin studies since, as we saw in chapter 2, the difference in percentage of genes shared between identical and fraternal twins (100 versus 50 percent) enables scientists to calculate heritability of traits.

One critic, biologist Marcus Feldman of Stanford University, points out that identical twins share many more external influences than fraternal twins: people tend to treat them alike, their parents dress them similarly, and they continue this trend when older. They often create their own private fantasy world to the exclusion of other people. Fraternal twins, on the other hand, typically behave no more alike than other siblings. "Whenever you measure heritability [using twin studies]," Feldman says, "you are glossing over the fact that the similarity of environments for identical twins is much greater than it is for fraternal twins."[9] Greater concordance of happiness levels between identical as opposed to fraternal twins may have at least as much to do with shared environment as with shared genes.

Feldman's account misses one fact, however: researchers look not only at identical twins reared together but also at those separated at birth and reared apart. These studies reinforce those conducted with nonsplit identical twins. Similarities between split identical twins can appear stunning, as David Lykken and his coworkers illustrate: "A pair of male [identical twins], at their first adult reunion, discovered that they both used Vademecum toothpaste, Canoe shaving lotion, Vitalis hair tonic, and Lucky Strike cigarettes. After that meeting, they exchanged birthday presents that crossed in the mail and proved to be identical choices made independently in separate cities."[10] Evidence like this deflates Feldman's objection.

A University of Pittsburgh team led by Bernie Devlin suggests that the environmental effect extends further back. Even twins separated at birth share an environment: the womb. Devlin's team believes that the uterine environment accounts for 20 percent of the IQ similarities between twins and 5 percent of the similarities between regular siblings, thus reducing the genetic effect.[11] The same theory presumably applies to other traits, such as happiness, but we need more research to support this possibility. Scientists disagree on how and to what extent life in the womb affects the person-to-be. Other questions arise. Does the womb constitute an environment in the sense that we usually understand the word? Can other (genetic?) factors counteract the uterine effect? Even with answers to these questions, people like Devlin would still believe that our genes (in combination with other factors) affect our behavioral traits. It would make little sense for people like Begley to quote his work as undermining a genetic basis for behavior.

Grant Steen, a medical researcher at the Saint Jude Children's Research Hospital in Memphis, expresses concern that scientists calculate heritability in one population group, say, white male twins, and simply apply these statistics to quite different groups, say, black male and female nontwins. Since different population groups experience different circumstances and life situations, he feels that automatic transferral of results must breed suspicion. Steen misses yet another important aspect of twin studies, namely, that they use a representative population sample: "Some of the twins did not reach the eighth grade," explain Lykken and Auke Tellegen, "whereas others have doctorates; they live on farms, in small towns, in big cities, and in foreign lands; their socio-economic levels are representative of Minnesota-born adults."[12] Even though Lykken and Tellegen's twins are predominantly white, psychologists David Myers and Ed Diener recount independent evidence showing that at least variations in levels of happiness don't depend on race or ethnic group.[13]

Lawrence Wright throws further doubt on twin studies in his book *Twins: Genes, Environment, and the Mystery of Human Identity*, in which he recounts the case of a pair of identical twins: one healthy, the other with a fatal version of the genetic disorder muscular dystrophy. John Burn, the doctor examining the case, concluded, "Even though [the twins] share the same genes, a genetic trait doesn't have to be shared."[14] We must exercise caution with twin studies. In the case of happiness, however, evidence from psychological studies of individuals as well as the twin studies supports the existence of a genetic set range. In addition, muscular dystrophy is a disease involving unequal distribution of faulty genes, whereas happiness is a normal trait of healthy human beings; it's not obvious that we can draw a useful parallel.

Scientists used to assume that identical twins were born in the same gestational sac and nonidenticals in different sacs. Wright points out that DNA testing has proved this reasoning faulty: roughly one-third of identical twins are born from separate placentas, and occasionally placentas belonging to fraternals merge. He concludes, "Many same-sex twins who believe that they are fraternal may actually be identical, and vice versa."[15] This observation by Wright doesn't sound against the twin studies on happiness. The bulk of the confusion concerns identicals who are really fraternals. Those fraternals mistakenly placed in the identical pool will *decrease* the degree to which identicals apparently attain the same level of happiness. Statistics for identicals attaining the same happiness levels are already high (44 percent as opposed to 8 percent for fraternals), and if some of these identicals are really fraternal, the statistics for identicals should rise above the 44 percent. We need ideally to compare the old results with new results employing the new criteria for distinguishing types of twinhood, but at the moment it looks as if even more support for the heritability of happiness would spring from this confusion.

Criticism of the studies of happiness using twins doesn't significantly impede their results.

THERE'S MORE TO US THAN A BUNCH OF GENES

Other critics concentrate on the "there's-more-to-us-than-a-bunch-of-genes" response. This stresses that behavioral geneticists like Hamer reduce the holistic human experience of happiness to nothing but the action of genes, electrical activity, and chemicals. These subjectivist critics push antireductionism, claiming that geneticists and their popularizers ignore the real subjective realm by confounding human experience with mere physical phenomena:

- Walter Freeman of the University of California at Berkeley, for example, says, "Joy comes with activities that we share with people we have learned to trust, and that enable us to share meaning across the existential barrier that separates each of us from all others. So happiness is not made by a chemical."[16] Scientists can stimulate our brains with electric currents, and we may report feeling pleasure, but happiness doesn't consist in this. Neither does it consist in the elation people feel after taking recreational drugs, like amphetamines or cocaine. "There is more to brain function," Freeman concludes, "than chemistry or electricity."[17]
- Writes Mark Epstein, "True happiness is the ability to receive pleasure without grasping and displeasure without condemning, confident in the knowledge that pain and disappointment can be tolerated."[18]
- "It's worse to wake up in the morning without having a larger purpose in life," says developmental psychologist Carol Ryff of the University of Wisconsin, "than to

wake up unhappy. Just feeling good is a poor measure of the quality of a person's life."[19] A truly satisfying life comprises a cornucopia of which the pursuit of happiness forms only a part.

- Behavioral genetics oversimplifies the reality.

Subjectivist responses such as these actually don't contradict behavioral genetics. Hamer's and his colleagues' work suggests only that genes provide a *percentage* of input into behavior. Hamer claims that 80 percent of our happiness in the long term depends on our genes. That still leaves 20 percent for other influences to make their mark, and we can also move up and down within the scope that our genes establish (after all, the 80 percent refers to a range). "Though genes may determine our average set [range] for happiness," Hamer writes, "they don't specify where we are within our individual range at any particular point in time."[20] Further, Plomin considers that genetic variation usually accounts for less than half the differences between the ways we behave.[21] Our moods, our minds, and our volition can all play their part. Subjectivist dismissals of biological explanations therefore go too far; genetic and subjective explanations may coexist.

Such statements as Freeman's, Epstein's, and Riff's suggest that joy and its correlates differ markedly from subjective well-being. True happiness is a loftier affair, a far nobler prize. Intellectual or spiritual satisfaction—whether through an intellectual search for the good, a struggle to fill our lives with virtuous thoughts and deeds, or a quest to lose our personal identity on the way to *Nirvana*—exceeds the biological so much that it becomes something else. One could of course just define true happiness or joy as having nothing to do with subjective well-being and thereby consider the problem solved. But that would do injustice to the words; a square doesn't become a circle just by wishing or defining it to be so. "True" happiness and joy must involve subjective well-being and therefore to some extent behavioral genetics. We can't divorce them from the physical.

LET'S HAVE 'EM BOTH

The "let's have-'em-both" response represents a more sophisticated version of subjectivist criticisms. Proponents admit the validity of results from behavioral genetics and other sciences, but they set up a dualism: on one side lies the mind with its feelings and sensations, and on the other lies the brain with its neurotransmitters. They create a unique spiritual status for happiness and thereby divorce it from the biological stuff of genes and brains. Up floats the ghost in the machine. The French philosopher René Descartes (1596–1650) wrote a book called *Meditations,* in which he advocates mind–body dualism. He famously claims, "Cogito ergo sum" (I think, therefore I am), meaning that, in essence, he comprises a thinking being (or mind) that doesn't depend on anything material (the body) for its existence. The world consists of two incompatible substances: mind or consciousness and matter. For Descartes, our brains and bodies form part of the material world and so don't participate in our essence as thinking beings.

Descartes soon realized, however, that mind and body are not as distinct as his theory implies because they constantly interact. When we fall and break a limb, our bodies suffer damage. We feel pain as a result of that damage. Physical events constantly produce mental effects. The same applies to happiness. When we build trusting relationships, perform virtuous deeds, or share sex with a special partner, we engage in activities that depend on our physical natures. We feel happy as a result of the experience. Physical events can produce happiness.

Happiness depends on both physical and spiritual aspects. Failure to acknowledge this closes spiritual thought off from the opportunities behavioral genetics offers; it also prevents spiritual traditions from offering insights to scientific investigations of happiness.

LET'S HOPE IT'LL ALL GO AWAY

A last-ditch response to scientific explanations—the "let's-hope-it'll-all-go-away" approach—employs avoidance tactics. Spiritual thinkers play a wait-and-see game: we will approach gene studies seriously only when and if someone else adequately answers all possible criticisms. These tactics look suspiciously like blind panic, akin to closing your eyes tightly at a busy road intersection and pumping on the gas pedal. While later research may modify the current results of the gene studies, the bulk and significance of the work will probably remain. It's a good bet that more and more evidence will mount to show a (partial) genetic and biological basis for such traits as human happiness. This will happen quicker than we think given the rate at which the Human Genome Project was completed.

CONCLUSION

We've reached a happiness stalemate. Neither scientists nor spiritual thinkers will give way. Neither recognizes the other's contributions as valid or even relevant. If they converse at all, they do so at cross-purposes. The result: voices raised in anger and an unhappy conflagration. This needn't be so. We've already seen how scientific and spiritual approaches to happiness might coexist in that the genetic set range still leaves room for the affects of other things such as mind, free will, and spirituality. Determinism doesn't necessarily rule the day. We need to build on, not deny, this compatibility. But what concessions must each side make? We also need to ask different and not defensive questions. Do the genes of the spiritually enlightened predispose them toward great happiness? How do genes and neurotransmitters relate to an afterlife? Trying to answer such questions may help us move beyond the conflagration. Scientists and spiritual thinkers need to stop their bickering and instead work together toward a richer, more meaningful account of the nature of happiness.

CHAPTER 3

Love and Spiritual Traditions

> Love is patient and kind; it is not jealous or conceited or proud; love is not ill-mannered or selfish or irritable; love does not keep a record of wrongs; love is not happy with evil, but is happy with the truth. Love never gives up; and its faith, hope, and patience never fail. . . . These three remain: faith, hope, and love; and the greatest of these is love.[1]

Love lies at the heart of our lives. It forms the basis of many human relationships: we love our partners, our children, our parents, and our close friends. The speed and generosity with which wealthy nations respond to urgent calls for famine aid for the starving in Somalia, Rwanda, or Sudan represents our love for the wider human circle. We even express love for places, literature, and music; a hauntingly beautiful landscape or an evocative aria can stir feelings of passion deep within us. Spiritual traditions the world over have added much to our understanding of this, a most enduring human emotion.

PLATO

Plato's *Symposium* describes a merry drinking party during which the participants eulogize and air their views on the nature of love (*eros*). Socrates (Plato's mouthpiece) describes love as a bridge between the divine and human worlds, "the medium of all prophecy and [spiritual practice], whether it concerns sacrifice, forms of worship, incantations, or any kind of divination or sorcery."[2] Love represents a hybrid: neither mortal nor immortal, neither continuous nor broken, but all at once.

Eternal beauty is good, according to Plato. We all desire beauty (and so goodness), and, in the act of loving, we reach out for that beauty. We can express love and attain beauty in different ways or on different levels. At the common or garden level, Romeo's love for the beautiful Juliet represents his desire to secure immortality through their offspring. More poetically, his love represents his passion to join with a kindred spirit and to generate sound moral principles and rules of life. This voices our common desire to find a partner who shares our values and concerns, with whom we can build a spiritually rewarding life. Love represents, at the most rarefied end of the spectrum, the scholar's endeavor to enrich the stock of human knowledge though study and dialogue. Plato frequently speaks of "giving birth" to philosophical or scientific ideas. The philosopher, as so often in Plato, holds the trump card: knowledge leads to true beauty, or goodness-in-itself, with all other forms of beauty but mere copies.

"Love must be desire for immortality," claims Plato.[3] In loving, we seek to attain goodness for ourselves, for eternity. The *Phaedrus* illustrates the intensity of the love that we can feel for another: we want nothing except to be by our lover's side, and we'll do anything, give up everything (friends, family, material possessions), to achieve that goal. This vision enchants us. Innumerable fairy-tale princesses have insisted that their paramour complete some dangerous deed in a far-away land just to prove the strength of his love—and the poor men invariably comply!

AUGUSTINE

Early Christian writers held differing views on the nature of love. Augustine defines the cardinal virtues of temperance, fortitude, justice, and prudence as forms of love. Each time we refrain from having just one more pint of beer, we express love. By confronting the ravages of cancer so bravely and cheerfully, Linda McCartney also expressed love. Ultimate love consists of a life lived by God's side, "whose eternity," says Augustine, "is true and whose truth is eternal, whose love is eternal and true."[4]

AQUINAS

"A thing is said to be loved, when the desire of the lover regards it as his [or her] good. The attitude of disposing of the appetite to anything so as to make it its good is called love. We love each thing inasmuch as it is our good."[5] So claims the medieval spiritual thinker Thomas Aquinas. It smacks of selfishness, however, to assert that degree of personal gain (or "good") governs the extent of our love. It also challenges the popular belief that loving involves personal *sacrifice*: we no longer have just ourselves to think about and must accept our lover's foibles with consideration and understanding. The words of Tom Lehrer's song express this:

> In winter the bedroom is one large ice cube, and she squeezes the toothpaste from the middle of the tube. Her hairs in the sink have driven me to drink, but she's my girl, she's my girl, she's my girl, and I love her.[6]

Aquinas steers a course around this difficulty by equating self-interest with the desire to improve humanity's lot. So he urges us to love God above everyone else (even the girl who squeezes the toothpaste from the middle of the tube) because God desires the common good of humanity (not just of specific human beings) and possesses the power to secure it.

CHRISTIANITY

Love lies at the heart of orthodox Christianity, as epitomized by Mark's account of Jesus' words:

> The first [commandment] is, "Hear, O Israel: the Lord our God is the only Lord; love the Lord your God with all your heart, with all your soul, with all your mind, and with all your strength." The second is this: "Love your neighbor as yourself." There is no other commandment greater than these.[7]

Jesus leaves his disciples with the command to "love one another as I have loved you."[8]

Love permeates Christianity in three forms: God's love for humanity, humanity's love for God, and human beings' love for one another. God's love for us takes the form of a creative, bestowing force, which culminates in the gift of the son, Jesus Christ, whose death and resurrection absolve us of our sins and enable us to live eternally with God. But love and God interconnect more closely than this: God's essential nature comprises love, and this love is available to everyone, everywhere. "Love is from God. Everyone who loves is a child of God and knows God," writes John.[9] The god of the New Testament represents a god of love.

God bestows love on us freely. By loving God and our fellow humans, we respond to God's love for us. We are also to treat grace and God's actions toward us as inspiration and examples in much the same way as children display love for their parents by following their lead and copying their behavior. We too must offer our love freely, without expectation of reward or return. Our love should extend way beyond friends and family, embracing even our enemies, because God loves each one of us. Matthew reports Jesus' command: "Love your enemies and pray for your persecutors; only so can you be children of your heavenly Father, who makes his sun rise on good and bad alike, and sends the rain on the honest and the dishonest."[10]

How about sexual or erotic love? New Testament writers tend to use the word *agape* to stand for love rather than *eros*. Perhaps they wanted to avoid the strong sexual overtones that *eros* carries in Greek literature.

Anders Nygren thought so. His book *Agape and Eros* emphasizes the sharp differences between the two terms, claiming they have nothing in common. He writes that, while *agape* describes a love involving God's response to us and our response to God, *eros* reflects human egocentricity and acquisitiveness, our ever-present desire to win an attractive sexual partner. Yet the Old Testament celebrates marriage and sexual love:

> I will sing the song of all songs to Solomon that he may smother me with kisses. Your love is more fragrant than wine, fragrant is the scent of your perfume, and your name is like perfume poured out; for this the maidens love you. Take me with you, and we will run together; bring me into your chamber, O king.[11]

So opens the Song of Songs in praise of erotic love. Further, Christian marriage is a sacrament in which each partner makes solemn vows of love, support, and fidelity to the other.

Sexual or romantic love need not be grasping or self-centered; it need not be inferior to altruistic *agape*. Any successful union requires sympathy, understanding, and give-and-take between partners. It should also involve romance, sex, and intimacy.

The love theme carries through to the most modern Christian communities. Who hasn't been stopped on the street by an evangelist touting an invitation to a prayer meeting or church service? We may find this intrusive and irritating, yet the evangelist views the action as an expression of love, an attempt to draw us into the fold and so to introduce us to an altogether better way of life.

"The greatest disease in the west today is not TB or leprosy," believes Mother Teresa, "it is being unwanted, unloved and uncared for. We can cure physical diseases with medicine, but the only cure for loneliness, despair and hopelessness is love."[12] By expressing her intense love for all people, Mother Teresa touched the lives of innumerable poor and sick people on the streets of Calcutta. She remains perhaps the most striking modern example of Christianity's number one ideal: love.

CONFUCIUS

"What you do not wish to yourself, do not do to others. Then neither in the country nor in the family will there be resentment against you."[13] So declares Confucius in his *Analects*. The family plays a strategic moral role within the ancient Chinese doctrine of Confucianism, as we saw in chapter 1, and this continues with the Confucian understanding of love. The Chinese term *hsiao* (one of several Chinese words referring to the idea of love) translates as "filial piety." A gentleman exhibits *hsiao*, according to Confucianism. He displays an affectionate, gentle nature that attracts similar people, and through relations with others, he promotes goodness. He's a kind of moral Superman. *Hsiao* also draws individual members of families, or clans, together. Confucianism dictates that family members should display a high degree of affection, consideration, and patience toward one another and a lesser degree toward outsiders—a concrete version of the old Western adage "Charity begins at home." Such good-heartedness (or *jen*) indicates a virtuous person—the Confucian gentleman—characterized by kindness, gentleness, compassion, and altruism. *Jen* enfolds goodness, wisdom, and courage, so much so that Confucius insists, "It is only the man of *jen* who knows how to love and how to hate people."[14] Only the person of *jen* possesses true moral insight and judgment.

MOHISM

The Mohist school of philosophy (originating in the fifth century B.C.E. with Mo-tzu) taught a different doctrine of love from the Confucian. Mohists believe that an excess of love for self, family, or state causes strife and social disorder. They instead advocate that human beings take the will of Heaven (*t'ien-chih*) as their example. We should love everyone equally, as does the will of Heaven:

> If all people regarded their parents, their elder sibling, and their ruler just as they do themselves, toward whom would they be lacking in devotion? . . . If everyone in the world practiced universal love . . . then the whole world would enjoy peace and perfect order.[15]

This doctrine of universal love bears a striking similarity to the Christian command to love our neighbors (even our enemies) as much as we love ourselves.

To express universal love is good in itself, according to Mo-tzu, plus it encourages good government and politics. A universally loving ruler ensures a peaceful and contented nation; the ruled reciprocate the love of their ruler. President Suharto of Indonesia would have done well to heed Mo-tzu's teachings. He preferred amassing businesses and capital for himself and his family rather than concentrating on the well-being and prosperity of his people and country. Suharto's egotism—a kind of Confucianism gone mad—fueled scenes of rebellion and pure hatred on the streets of Jakarta.

TAOISM

The ancient Chinese school of Taoism also says something about love. People who follow the spirit of the Tao (or the Way) by living in perfect harmony with the forces of nature personify *wu-wei*, according to the sixth-century B.C.E. text the *Tao-te ching* ("The Way and Its Power"). *Wu-wei* represents the Taoist ideal of inaction (or lack of striving and aggression), as illustrated by the proverbial sage who "does nothing, yet accomplishes everything." Someone who has attained the perfection of the Way lives according to the virtues of love, compassion, mercy, patience, meekness, tenderness, and generosity toward all living beings. Love, for the Taoist, results from a detachment from the striving and concerns of everyday existence; it arises from a life of peaceful meditation. This supplies the logic for the Taoist notion that love has nothing to do with emotion or passion. "The blankness of the Unnamed brings dispassion; to be dispassionate is to be still. And so, of itself, the whole empire will be at rest."[16]

BUDDHISM

The legendary figure of the *bodhisattva* ("he whose essence is enlightenment") features heavily in Mahayana Buddhism. The *bodhisattva* made a promise:

> However innumerable the beings, I vow to save them. However inexhaustible the defilements, I vow to extinguish them. However immeasurable the teachings, I vow to master them. However incomparable enlightenment, I vow to attain it.[17]

The promise summarizes the *bodhisattva*'s decision to reject the peace of *Nirvana* (the final release from the cycle of death and rebirth) until he has guided all other beings to that state. This involves more than mere shepherding; the *bodhisattva* participates in our suffering, shouldering many of our woes. This decision springs from the infinite love and concern that the *bodhisattva* feels toward all other beings. He thus enfolds a host of virtues, including infinite compassion, giving, morality, patience, vigor, and meditation. He shows neither concern for nor consciousness of his own self.

The *bodhisattva*'s love and compassion share similarities with the Christian god's love for humanity. Both concern themselves only with the well-being of others. Both fail to show any self-interest; they expect no reward for their actions. Both have made enormous personal sacrifices (the rejection of *Nirvana* and the gift of a son) for the sake of others.

Mahayana Buddhists believe that a number of different *bodhisattvas* exist, each representing an aspect of the Buddha. They consider the Buddha the supreme *bodhisattva*: he can bring our suffering to an end and facilitate our entrance into *Nirvana*. He visits the Buddhist hells to ease suffering and to point the way toward salvation. Artists capture this aspect of the Buddha's nature by depicting him with a thousand eyes and arms all working together to help him seek out and assuage our suffering.

HINDUISM

Love plays a multifarious role in the Hindu tradition. The word *kama* denotes pleasurable experiences, especially erotic ones. Kama also names the Indian god of love, who embodies desire, sexual pleasure, and romantic union, and who identifies with the life energy that courses through all living creatures.

The Vedas (the earliest Indian Scriptures) tell of the Vedic sacrifice, which involved showering the gods with adulation and gifts of food and drink. The gods, in return, provided the people with fertile lands, health, and wealth. This story treats love as reciprocity: the people's giving facilitates the gods' giving. Love involves the mutual exchange of gifts.

The *Bhagavadgita* (the Indian equivalent of the New Testament) raises the understanding of love from the erotic or practical to the spiritual. It names Krsna as the one and only personal God. By exclusively loving and worshipping Krsna, we attain salvation. By focusing on Krsna alone and rejecting other spiritual traditions, we come to know God truly and become one with God. Krsna returns the love of the true devotee and, through divine grace, brings an end to the cycle of karmic rebirth (or retribution for our misdeeds in previous lives). We ensure eternal happiness and peace if we devote ourselves to God. How do we do this? The true devotee, Krsna explains, "does my work, . . . is devoted to me and loves me, . . . is free from attachment [to worldly values] and from animosity toward any creature who comes to me."[18] We must exercise self-discipline, disregard both pleasure and pain, treat friends and enemies equally, and show concern and amiability toward all beings.

The god Kama represents romantic love and sexual union, while the Vedic depiction of the people's relationship with the gods involves human neediness and the struggle for survival. The kind of personal loving relationship with Krsna that the *Bhagavadgita* recommends parallels the Christian relationship with the loving God. By displaying love and concern for both God and their fellow creatures, devotees of Krsna embrace *agape*, as do Christians who love God and embrace friends and enemies alike. The split between erotic/practical and spiritual love in the Hindu tradition shares common ground with the Christian distinction (at least as Nygren describes it) between *eros* and *agape*.

ISLAM

Islam, like many other spiritual traditions, speaks of erotic or sexual love. Unlike Christianity, however, Islam highly praises sexual love: Muslims consider erotic pleasure a Divine Mercy, even a sacrament, not just an unfortunate consequence of procreation. Explains one Muslim author,

> When a man looks at his wife and she looks at him, God looks upon them both with mercy. When the husband takes the wife's hand and she takes his hand, their sins vanish between their fingers. . . . Pleasure and desire have the beauty of mountains.[19]

Earthly love even moved the Prophet Muhammad to spiritual inspiration. When he saw and fell in love with Zaynab (one of his wives-to-be), he declared, "O Overturner of Hearts, make mine steadfast."[20] Sexual pleasure gives us a taste of the life to come in *Al-janna* (the garden), where both spiritual and sensual happiness reach their zenith.

Muhammad didn't encourage celibacy or abstinence from marital relations. "I fast, and I break the fast, I pray, I sleep, I go in unto women; beware! Whosoever deviates from my *sunnah* [example] is not among my followers."[21] So declares one Hadith (report of Muhammad's words and deeds). Muslim marriages must be consummated within forty days; otherwise, they become void. *Qadis* (or judges) may even berate husbands and wives who fail to fulfill their duties toward their partners.

Sexual love verges on the holy for Muslims and brings the participants nearer to God as well as to each other. Enjoyment of sexual pleasure counts as a virtue. Teacher and mystic Ibn 'Arabi neatly captures the sentiment: "The most intense and perfect contemplation of God is through women, and the most intense union [in the world] is the conjugal act."[22]

JUDAISM

Love is the core of God's being. For the Jewish tradition, God's love of us is the primary instance of love. It is so strong that it can ward off the severity of the divine retribution to which we open ourselves by falling far short of the standards God has set for our behavior.

The most important love we can have is for God. "The highest principle of religion and . . . the ideal of human perfection," writes Jewish scholar K. Kohler, "is attained in love of God such as can renounce cheerfully all the boons of life and undergo the bitterest woe without a murmur."[23]

Our love for other people is our attempt to emulate God's love for us. The love we ought to show to each other is, in the Hebrew Scriptures, *chesed,* probably best translated as "faithful love." Judaism perceives the love command differently than does Christianity. We can't carry out the Golden Rule, "Love your neighbor as yourself," literally. "We cannot love the stranger as we love ourselves or our kin," writes Kohler. "Still less can we love our enemy." Hebrew Scriptures teach that we can and ought to treat our enemies magnanimously and forgive them but that we can't truly love them unless they turn from enemies to friends. Love by itself doesn't suffice because "it is too much swayed by impulse and emotion and is often too partial." Justice (maintaining the distinction between right and wrong) is also needed. "Love without justice leads to abuse and wrong."[24] A harmonious combination of love and justice motivates our right conduct toward others. "Justice requires what is due to others by virtue of their humanity," writes Rabbi Israel Mattuck, "love serves them to satisfy the needs of their humanity."[25]

CONCLUSION

Love intertwines itself deeply through the major spiritual traditions, both Eastern and Western, whether in the form of our love for one another (both *agape* and *eros*), our love for God, or God's love for us. Love plays a central role in many of these spiritual traditions: the *bodhisattva* out of love shares our suffering with us, Krsna releases truly loving devotees from the cycle of karmic rebirth, and Jesus Christ loves us so much that he died to absolve us of our sins.

We as well need love: to be loved brings feelings of well-being and security, to love brings a sense of achievement and direction. Many of us feel that a loving relationship makes our lives complete.

We might expect the spiritual traditions to speak of love in spiritual terms and that this includes the emotions: our feelings toward one another and God's concern for us. Is there another side to love? Is love, like happiness, rooted in our biology? Loving, after all, takes the form of concrete, physical activity: we feed and clothe the child we love, the *bodhisattva* assuages our suffering and points the way toward salvation, and sexual love, above all, involves intense physical intimacy. Loving feelings translate into loving actions. Is love rooted in more than our spirit and emotions?

CHAPTER 4

Love and Contemporary Science

> Oxytocin is a marvelous molecule, influencing our lives through touch. . . . [It] bonds and attaches us to those we love—or perhaps causes us to love those it bonds us with.[1]

A child in school steals, cheats, fights, and lies. No matter what adults try, they cannot turn him into a responsible and loving boy. Teachers blame his family background; parents call for a special educational program; counselors work on building self-esteem. The established system considers outside intervention the cure. Does the problem lie outside or inside the boy?

The scientific search is on:

- "A natural chemical called oxytocin is found to underlie love," writes Robert Wright.[2]
- In his book, *Living with Our Genes*, Dean Hamer explains, "Everyone, gay or straight, feels the tug of genes involved with sex and love, from the sharp pangs

of puberty, to the defining role of gender, and the fierce, protective feelings of a parent for a child."[3]

- A group of neuroscientists, social and behavioral scientists, clinical psychologists, biological psychiatrists, and psychoanalysts gathered at a symposium—like several other recent gatherings—to discuss the question, "Is there a neurobiology of love?"

Love emerges from the emotional closet and scientists struggle to isolate and understand the physiological processes underlying our expressions of romance, responsibility, and caring.

OXYTOCIN AND VASOPRESSIN

Inside our brains lies a hypothalamus, the organ that controls "primitive" behaviors such as sex, aggression, and feeding. It produces the hormones oxytocin and vasopressin, which then pass through a stalk down to the posterior pituitary gland at the base of the brain for storage and secretion. Both of these biochemicals evolved from the primordial hormone vasatocin, which still endures in the lower vertebrates such as fish. Both share a similar molecular structure, differing in only two of their nine amino acids, the building blocks of proteins. When released into the body, they bind to specific targets called receptors located in the brain and elsewhere, like keys fitting into locks. The receptors then affect other body parts and, finally, behavior.

Medicine has long known the effects of oxytocin on the female reproductive system. It naturally stimulates contractions in the uterus right through the birth process by locking onto specific receptors in the muscles of the uterus, causing them to tighten. Not surprisingly, the name "oxytocin" derives from the Greek for "swift birth." Obstetricians inject a synthetic form of it to arouse contractions when labor flags. (The Nobel Prize–winning work of Vincent du Vigneaud in

1953 pioneered its and vasopressin's synthesis.) It also helps control excessive bleeding after delivery of the infant. It prompts the mother's mammary glands to release milk within seconds after her baby begins to suckle. So susceptible is oxytocin release to emotional influences that even the cry of a hungry baby can prematurely stimulate milk letdown—as every mother and observant father knows. Oxytocin also plays a physiological role in highly emotionally charged activities like coitus, nipple eroticism, and female sexual responsiveness.

Research continues into the source and functions of oxytocin. Neuroendocrinologist Hans Zingg of McGill University led a study on oxytocin's role in prompting labor in animals. His team's report states that the uterus, rather than the hypothalamus, produces oxytocin for this purpose. Similar evidence occurs in humans: women who commence labor naturally have much higher placental concentrations of RNA (which governs production of oxytocin) than do women who give birth by caesarean section before natural labor begins, according to obstetrician Brian Mitchell of the University of Alberta in Edmonton.

Maurice Manning and his colleagues in the Department of Biochemistry at the Medical College of Ohio in Toledo developed several synthetic oxytocin "antagonists." These copycat proteins chemically resemble oxytocin and bind to the appropriate receptors, blocking out the real oxytocin molecules and thus the effects normally induced by the receptors. For those mothers who experience premature labor, injecting an oxytocin antagonist can suppress contractions. "There's no question that the oxytocin antagonist will knock out the contractions—it's really good at that," comments reproductive physiologist Laird Wilson of the University of Illinois, Chicago.[4]

Oxytocin's companion hormone, vasopressin, also plays a key role: it maintains a constant volume of water in our bodies and regulates to within narrow limits the concentration of dissolved substances in the fluids outside our body cells. Many specifically male functions of certain animals—such as marking territorial boundaries with scent—involve vasopressin, giving it a reputation as the "real man's" molecule.

Recent research with vasopressin and oxytocin begins to inculcate fields other than medicine. Oxytocin receptors may play a much larger role in *social* behavior than researchers previously thought. Peter Klopfer of Duke University links the formation of social bonds between mothers and their children to the release of oxytocin. Could it be the hormone of "motherly love"? he wonders. Physiological changes might prepare an expectant mother, arming her with the psychological tools she needs for her unfamiliar new role. Niles Newton of Northwestern University speculates that the secretion of oxytocin influences our sexual as well as our maternal bonds.

The research gains momentum. A virgin female rat in the laboratory, when first presented with pups, usually ignores them, is frightened of them, or eats them. She will tolerate them only when they are introduced to her many times and over several days. Then she may even care for the youngsters by licking them, retrieving them when they stray from her side, and crouching over them protectively—just what we expect from a responsible mother. A pregnant rat, on the other hand, responds to pups caringly within minutes, even before delivery of her own. In a famous experiment from 1968, Joseph Terkel and Jay Rosenblatt of Rutgers University injected virgin female rats with blood from rats that had just given birth. The result: the virgins nurtured the pups in significantly less time. Could something in the blood elicit a maternal response? Could that something be oxytocin? Further, parent rats can mistreat their children when injected with antagonists to block their oxytocin receptors, according to research by Cort Pedersen and Jack Caldwell at the University of North Carolina and Gustav Jirikowski at Scripps Research Laboratory.

VOLE RESEARCH

Voles are small, brown, nondescript mammals of the genus *Microtus* that live under seeds and grasses. Members of one species—the prairie vole—put us humans to shame. They share elaborate systems of burrows and feeding tunnels, and, unlike most rodents, males and females form long-lasting bonds, raising their young together. On the other hand, montane (or mountain) voles are home wreckers par excellence. They occupy separate burrows and avoid each other except to mate—which they do often and indiscriminately. Mother montane voles usually abandon their pups sixteen days after birth and fathers never see their offspring. When a predator plucks a youngster from the nest, it neither calls for help nor surges with stress-related hormones. Why, in comparison with their prairie cousins, do the high-meadow montane voles lack family values and behave so asocially?

"Do these species differ with respect to central pathways for oxytocin or vasopressin?" asks Thomas Insel of Emory University (coauthor, with Sue Carter of the University of Maryland, of many vole studies). He replies, "[The species] differ in the neural distribution of receptors for both [hormones] as much as they differ in behavior."[5] Prairie voles have three times the number of oxytocin receptors in their prelimbic cortex—and seven times more in their nucleus accumbens—than do montane voles, according to research by Insel and Larry Shapiro, then at the National Institute of Mental Health's (NIMH) facility in Poolesville, Maryland. Yet receptors for other unrelated hormones match across the species. Might the differences in receptor distribution relate to the differences in social behavior?

Making this even more interesting is the fact that, during the brief period in which female montane voles nurse their young, the number of their oxytocin receptors surges, matching those of the female prairie vole. Could distribution patterns of oxytocin receptors account for social characteristics of monogamy and childcare? Insel thinks so: "This is evidence that oxytocin receptors may be very important for the so-called 'affiliative' behaviors that make animals receptive to social attachments," he asserts.[6]

As soon as the female prairie vole becomes sexually active, she and a male will copulate repeatedly, more than fifty times in over thirty-six to forty-eight hours. She becomes much more socially exclusive after this sexual frenzy, preferring her mate to unfamiliar males. This discovery by Diane Witt of the University of Maryland suggests that mating instills long-term pair bonding. Copulation causes the release of oxytocin in a number of small mammals; might this be the critical factor in developing the female prairie vole's social preferences and fidelity? Jessie Williams, also of the University of Maryland, found that a female prairie vole rapidly forms a preference for a male if exposed to oxytocin for six hours. However, when administered with an antagonist to block the oxytocin receptors, oxytocin ceases its social effect. "These results suggest that oxytocin's action within the brain may be one of the physiological events that lead to the formation of monogamous pairs," comment Carter and Lowell Getz of the University of Illinois.[7]

OTHER RESEARCH WITH MOTHERS AND OFFSPRING

Studies of domestic sheep by Barry Keverne, Keith Kendrick, and their colleagues at the University of Cambridge strengthen the case for oxytocin's social role. As a lamb moves down its mother's birth canal, it stimulates nerves that trigger the release of oxytocin. Only with oxytocin present at birth or injected so that it reaches the brain at the same time as mother ewe meets her newborn will she bond with her offspring. High levels of oxytocin also occur in her milk. She rejects her lamb if something blocks oxytocin's release. Perhaps the lamb's oxytocin ingestion helps forge a mutual attachment.

Further evidence for a link between oxytocin and mother–infant attachment emerges. Kerstin Uvnäs-Moberg of the Karolinska Institute in Stockholm, Sweden, tells us that a human infant's suckling leads to an oxytocin-mediated increase in blood flowing to the skin covering the mother's mammary gland. "A warm nipple may encourage attachment by the offspring," she explains. The warmth may "induce physiological effects of a calming and nurturing nature."[8]

Psychologists Eric Nelson and Jaak Panskepp experimented with fifteen-day-old rats and their mothers. When they associate an odor with their mother, the baby rats approach the smell more quickly and spend more time with it than when they fail to make a maternal association, provided they make the association after the administration of oxytocin. Administration of an oxytocin antagonist prior to association fails to produce the attachment effect.

More recently, however, Insel and his colleagues uncovered data that seem to fly in the face of the growing results about oxytocin. They created genetically deficient "knockout" mice that lack a working gene for oxytocin. Yet these mice can still mate, give birth, and display normal maternal behavior. "There's no question the females are fully maternal," confirms Insel. "It's also startling to find out that mice that have no oxytocin whatsoever seem to have pretty normal reproductive behavior. I found it very difficult to accept at first."[9]

Knockout mice do differ in significant ways from normal mice. For a start, they can't nurse their young unless artificially injected with oxytocin. "The milk is there. They just don't let it down in response to suckling," explains Scott Young of the

NIMH in Bethesda, Maryland.[10] They also display differences in social behavior, "mostly centering around aggression and social investigation," says Insel. "It appears the knockout mice don't investigate other mice as much."[11] Knockout pups also demonstrate a decreased tendency to emit ultrasonic isolation calls when separated from their mother and littermates. Perhaps these pups "fail to form social attachments early in life, and are therefore not distressed by the separation," speculate Insel and his coworkers.[12] Whatever the story with physiology, exposure to oxytocin still seems to influence social behavior.

Perhaps it shouldn't surprise us that oxytocin-deficient mice can still reproduce and care for their young. "Reproduction is just too important to have one mechanism for ensuring parturition [giving birth]," according to Louis J. Muglia of the Washington University School of Medicine in St. Louis. "There's bound to be redundancy in the system."[13]

Insel's team, however, offers a different explanation. The evidence derived from knockout mice contradicts several rat (and sheep) studies that say oxytocin does induce maternal behavior. But rats and mice differ importantly. Virgin *rats* do not display maternal behavior in that they ignore pups and can commit infanticide, yet, just before birth, a striking shift occurs: they become driven, relentlessly building nests and retrieving, licking, and protecting pups. Infusions of oxytocin facilitate the shift, but the administration of oxytocin receptor antagonists blocks it. Virgin female laboratory *mice*, by contrast, exhibit full maternal behavior on first exposure to pups. "Since there is no shift in maternal behavior which occurs at parturition in laboratory mice, it is not surprising that oxytocin-deficient mice show normal maternal behavior," Insel's team concludes. "We must be careful in making generalizations regarding the relationship between oxytocin and specific behaviors."[14] This kind of explanation suggests that oxytocin induces maternal care in certain species but may control milk ejection in many more. "The role of oxytocin in the regulation of social behaviors must be considered on a species-by-species basis."[15]

RESEARCH WITH MALES

What about males' sexual and parental behavior? After the initial sexual bout, a male *prairie* vole prefers his mate and ferociously guards against rivals, even in her absence. James Winslow, while at the NIMH in Poolesville, found that a male isolated from females and injected with vasopressin becomes aggressive and attacks other males. (Females respond little to vasopressin.) If exposed to a female and injected with vasopressin, a male develops a preference for her even if they do not mate. Administering a vasopressin antagonist to a male ready to mate doesn't prevent repeated and intense copulation, though afterward he doesn't fend off intruders or prefer his partner. An oxytocin antagonist, on the other hand, alters neither the male's mate preference nor his guarding behavior. Zuo-Xin Wang, Craig Ferris, and Geert De Vries of the University of Massachusetts in Amherst found that vasopressin also increases (and a vasopressin antagonist decreases) the amount of time a male spends with his pups, which he would typically and naturally do by retrieving them and huddling over them.

Administering vasopressin to the polygynous, nonparental male *montane* vole induces quite different effects. Larry Young and his colleagues at the Emory School of Medicine in Atlanta, Georgia, have shown that it increases his self-grooming but has no effect on his behavior toward either his mate or his offspring. The two species appear to express love in different ways: the prairie vole directs affection toward his family, while the montane vole directs it toward himself. "Receptors in the lateral serum, found only in the montane vole, might be responsible for the effects of . . . vasopressin on self-grooming," comments Insel.[16] Difference in distribution of vasopressin receptors between the two species may again account for the difference in social behavior.

Many biologists believe that the kind of paternal care exhibited by the prairie vole anchors male monogamy. Many believe that vasopressin (in conjunction with its receptors) plays a part in male sexual and parental behavior. "Vasopressin may have a more general role as a neuropeptide involved in eliciting parental care and defensive behavior with respect to self and family," suggest Carter and Getz.[17] "Vasopressin," adds Insel, "may . . . be sufficient for male pair bonding."[18]

While oxytocin encourages social contact, vasopressin compels the male's antisocial, guarding behavior after copulation. These hormones—their behavioral and cellular functions—counter each other in some circumstances. Perhaps oxytocin blocks the unfriendliness induced by vasopressin. On the other hand, no matter what the mechanisms, vasopressin and oxytocin *together* help determine for many species if a pair bond, nurture and care for their young, and defend their family. Vasopressin and oxytocin seem crucial for pair bonding, eliciting parental care and nurturing, and defending the family. Monogamy and polygyny therefore appear to express the net outcome of what happens when oxytocin and vasopressin activate different circuits in the brain.

HUMANS

We're all dying to know about humans. Does the love we display toward our partners and kids translate into talk about neurochemicals?

The human brain manufactures vasopressin and oxytocin molecules that bind to receptors there. So they exist and work in our brains. Our forebrain, in particular, contains many oxytocin receptors. Further, though vasopressin and oxytocin are large molecules and do not readily penetrate the blood–brain barrier, they exist in larger-than-normal quantities in the brain when hard at work elsewhere in the body. They could, in principle, influence our social behavior.

Much of the maternal behavior displayed by an expectant mother arises from hormonal changes that her system induces. After birth, it stems from sensory stimulation and interaction with her child. A sensitive period for mother–infant bonding occurs just an hour after birth, when the mother's oxytocin level rises markedly. Could oxytocin act as a catalyst for the bonding mechanism?

"Lactation is also accompanied by behavioral changes which may be linked to central actions of oxytocin," claims Uvnäs-Moberg.[19] Not only do women report feeling relaxed and sedated during nursing, but they also feel calmer and more socially interactive than do nonpregnant, non-breastfeeding women of a similar age. Uvnäs-Moberg found that the women's degree of calm correlates with their oxytocin levels. Further, women who give birth via caesarean section don't report the kind of personality changes noted by those who give birth naturally. Could the oxytocin released at (natural) birth and feeding enhance social and maternal feelings and so facilitate bonding between mother and child?

A British study of male medical students shows that vasopressin peaks in the bloodstream during sexual arousal and oxytocin at orgasm. In both female and male humans, oxytocin levels rise dramatically during sex. Perhaps this promotes the feelings of romance and infatuation that we associate with lovemaking.

We do share, it therefore seems, many *physiological* responses with other animals, but the important question is whether we share similar *social* responses and whether oxytocin and vasopressin promote them. "Do vasopressin and oxytocin provide the chemistry for human attachment?"[20]

Researchers exercise caution. Insel points out that although receptors for both oxytocin and vasopressin lie in the human brain, their distribution pattern resembles neither that of the monogamous prairie vole nor that of the promiscuous montane vole. Perhaps we humans reside in a halfway house between monogamy and promiscuity. Few of us today settle down with our first sexual partner, and even when in a stable relationship, our eyes rove and we appreciate the attractiveness of others. Insel stresses, however, that studies with nonhuman primates (squirrel monkeys) show that increases of oxytocin and vasopressin in the brain do influence social interaction.

"We know very little about the hormonal basis of social attachment in our own species," explain Insel and Carter. "In large-brained primates like us, the effects of oxytocin and vasopressin . . . are undoubtedly mediated—and perhaps inhibited—by many other factors, especially by the complex activities of our enormous cerebral cortex."[21]

CONCLUSION

It's tempting to dub oxytocin the "love molecule," but objectively establishing a causal association between oxytocin and love proves difficult. Too many variables confront the research. The roles of vasopressin and oxytocin are also difficult to document, even in animals. We clearly need to exercise caution when extrapolating data derived from animal or limited human studies. Many questions remain unexplored or are only partly explored.

Even so, the research does show a biological rootedness to love for humans as well as animals, whatever else it may involve. Social bonds possess a biology. Love is in part a physical trait derived from evolution. Parental, filial, and sexual love happen with oxytocin and vasopressin, which promote the behaviors and symptoms of loving. When people love this way, these chemicals occur in their bodies in larger-than-normal amounts; love functions with them.

How do these scientific findings relate to the spiritual love the previous chapter described? According to most spiritual traditions, we possess the power to choose whether to love. Jesus urged us to love our neighbors. Taoists strive to attain the perfection of the Way and so to display love toward all living beings. Yet we sometimes love despite ourselves; not many parents remain angry for long with their practical joker offspring. How much is free will involved in loving? Can we decide to love and do so even if this means the decision prompts the release of hormones? This asks a crucial question. The involuntary release of oxytocin serves well the survival of our offspring and with them our genes. Could such vital functions depend on our whims alone?

What of divine love? Do oxytocin and vasopressin drive the Divine's love and concern for us? Must the Divine embody hormones and biochemicals in the same way as do human beings?

Only with answers to the likes of these questions can the reconciliation of scientific and spiritual understandings of love commence.

INTERLUDE 2

An Unloving Conflagration

> Bonding comes . . . with shared activity . . . in which people learn about each other through cooperation. Knowing another person doesn't come with foreplay and orgasm. It comes in cooperative activities during and afterward.[1]

The past two chapters have painted starkly contrasting pictures of the nature of love. Spiritual thought teaches that love typifies the relationship between humanity and the Divine. Many belief systems glorify the power of love and urge us to love one another. Recent science, on the other hand, suggests that love comprises a cocktail of the hormones oxytocin and vasopressin and arises from the drive to pass on our genes. These two conflicting pictures threaten to collide.

DIFFERENCES

The science and the spirituality of love differ, like happiness, in terms of *definitions*. Love, from the spiritual point of view, embraces a moral dimension. Spiritual traditions advocate that we love both the Divine and one another. To express love is to live the moral life, and in many traditions, this secures happiness in the life to come. Even sexual or romantic love reflects how we ought to behave: Christianity teaches that we should love, care for, and remain faithful to our spouses, while Islam treats sexual love as something almost holy, an activity that brings us closer to God as well as each other. By contrast, love, from the scientific point of view, represents not what we ought to do but what we do do. It has its roots in biology rather than morality. The production and release of vasopressin and oxytocin govern our loving behavior toward our partners and our children. Love, for the scientist, is a matter of biochemistry.

The science and spirituality of love also disagree over the *will*. Many spiritual traditions embody commands to love the Divine as well as our family, friends, and neighbors. Such commands invoke our autonomy: they suggest that we can choose to love and then act on that choice. Scientific research implies, in contrast, the involuntary nature of our loving actions: when mothers give birth, their veins rush with oxytocin, and they bond with their babies; when we make love, hormones infuse our bodies, and we will go to the ends of the earth for our partners. We can't help our actions. As *Cosmopolitan* puts it, "When you fall in love or in lust, it isn't merely an emotional event. Your hormones get in bed with you too. . . . It's as if a corporate decision is going on here, with each chemical casting its vote. You may just find yourself outnumbered."[2] We act purely from our biology.

Finally, and perhaps most contentiously, some spiritual traditions emphasize that the Divine *is* love. "Everyone who loves is a child of God and knows God, but the unloving know nothing of God. For God is love." So proclaims John in his epistle.[3] Kama (the Indian god of love) embodies desire, sexual pleasure, and romantic union; Kama *is* (romantic) love. Science, of course, makes no mention of the Divine as love or of the Divine's love for us. This makes sense from the scientific point of view. We envisage the Divine existing as a spiritual (not a material) being. Applying biological concepts to nonbiological entities makes no sense.

REACTIONS FROM SCIENCE

Almost no one has noticed the potential problem despite the obvious discrepancies between scientific and spiritual accounts of the nature of love. Scientists continue their research apace, while spiritual thinkers still plough their own furrow. Neither shows much concern for—or even knowledge of—the other's endeavors. Perhaps we can excuse this because the neurobiology of love comprises a novel area of research, its ideas still very much in their infancy. Unlike the genetics of happiness, it has received little media attention, and the general public remains unaware of its existence. Why is this? Perhaps the media understand our psychology. We are used to distinguishing between love and lust. Lust we associate with drives and desires—the biological part of our being—and love we understand as part of our higher emotional or moral being. Perhaps this perceived distinction shields us from anxieties that might arise from familiarity with the biochemistry of love. The neurobiological underpinning of love therefore lacks mass appeal and so fails to make a good story.

At base, though, the issue lies more fundamental than that. Scientists and spiritual thinkers should show awareness of one another's work and they should actively engage with problems of mutual interest regardless of media interest and public concern. Their failure to do so derives from age-old (and continuing) antagonism, succinctly displayed in the seventeenth century by a Catholic cardinal: "The intention of [the Bible] is to teach us how one goes to heaven, not how the heavens go."[4]

A handful of science writers display some awareness that the recent work on love threatens to displace older spiritual ideas. "We must dig into our biology," says Marc Zabludoff in a *Discover* magazine editorial, "into the anatomic and chemical structure of our brains, into the microbes that invade our cells, and ultimately into the deepest, smallest part of ourselves, the beadlike bits of DNA that, strung together, make up our genes. These are the agents, we are told, that turn us toward joy or depression, love or murderous rage, dominance or submission."[5] Explains Robert Wright in his *Time* cover story, "The most ethereal parts of life—the things that once seemed heaven-sent—have fallen steadily within the reach of concrete explanation. The mapping of our finer feelings to neurotransmitters and other chemicals proceeds apace. Love itself—the love of mother for child, husband for wife, sibling for sibling—may boil down, in large part, to a chemical called oxytocin."[6]

Neither Zabludoff nor Wright expresses concern, though, that the science of love might oust spiritual thinking, nor do they indicate that we might gain a more complete understanding of love through the reconciliation of scientific and spiritual doctrine. "Does such [a scientific] explanation suffice?" asks Zabludoff. "Certainly it doesn't move us as does the *Iliad* or the Psalms of David. But what we've lost in metaphoric narrative," he answers, "it would seem we've gained in knowledge. Neurotransmitters and proteins and brain structures may not be the stuff of myth and epic poetry, or even of a halfway decent action movie, but they are most assuredly the stuff of science. And science . . . is the foundation for our present and future understanding of the world."[7] Wright displays a similarly cavalier attitude: "It seems somehow harder to rhapsodize about the universal love many [spiritual traditions] prescribe when you know that, if it ever comes, it will rest on the same stuff researchers inject into rats to make them cuddle. Another bit of less-than-inspiring news is the clearer, more cynical, understanding of why love exists—how it was designed by evolution for only one discernible purpose: to spread the genes of the person doing the loving."[8]

Spiritual doctrines of love represent, from the scientific perspective, nothing more that a pleasing story, a touch of gilt on top of the skyscraper of knowledge. For a full understanding of the meaning and nature of love, we need look no further than the science's explanations. Spiritual thought must stand aside and assume its proper place.

REACTIONS FROM SPIRITUAL TRADITIONS

What about the reaction of spiritual thinkers to scientific findings about love? There appears to be no reaction; spiritual thinkers either remain unaware of the research or persist in turning a blind eye, confident that their position emerges unscathed.

If spiritual thinkers were to react to the neurobiology of love, their responses would probably follow their strategies for combating the genetic basis of happiness, which interlude 1 discussed. Perhaps such responses will emerge if the media tide turns and attention focuses on the biochemical basis of love.

LET'S-GET-'EM

How might the specific arguments run? The "let's-get-'em" response would involve a direct and sophisticated assault on the methods and results of neurobiology and biochemistry. One possible attack might target the fact that, so far, studies concerning the link between oxytocin and vasopressin remain limited. Science can't conclusively confirm that oxytocin and vasopressin together induce parental behavior and long-term pair bonding. Nor can it explain in detail the process by which hormones influence behavior. While some truth lies in this objection, more and more evidence amasses to defeat such claims. Scientists continually publish new results that confirm the hormonal–behavioral link, and these studies correlate over a number of species: montane voles, prairie voles, sheep, rats, squirrel monkeys, pine voles, and meadow voles. Explanation via the biological processes becomes easier with the accumulation of results. Evidence for a neurobiology of love won't disappear whatever else happens.

Proponents of the "let's-get-'em" response might now change tack, arguing that, while evidence may support a neurobiology of animal love, this doesn't justify extrapolation to the human case. Of course, we must exercise caution when moving from animal to human studies, but the move does have some evidence on its side. Oxytocin and vasopressin do act within our bodies, as we know from the previous chapter, their release coinciding with certain social behaviors. Neither should it surprise us that both animals and humans display similar physiological and social responses. Attachment can take three forms, according to Thomas Insel: parent-infant, filial, and male-female. "None of these forms of attachment is uniquely human," he goes on to point out, "suggesting that the neural basis can be investigated in animal models."[9] Increased transmission of oxytocin and vasopressin in one of our closer animal relatives—the squirrel monkey—influences social behavior. It thus becomes more and more likely that we react in similar ways. Insisting otherwise smacks of an unacceptable anthropocentricity.

THERE'S-MORE-TO-US-THAN-A-BUNCH-OF-BIOCHEMICALS

Adherents of the "there's-more-to-us-than-a-bunch-of-biochemicals" response might insist that, though loving involves hormones and biochemicals, love itself comprises much more. We can't describe love solely in terms of chemical analysis since human experiences, emotions, and feelings don't reduce to physical phenomena. Walter Freeman of the University of California at Berkeley again flies the subjectivist flag: "Trust emerges not just with sex," he says, "but also with vigorous shared activity in sports, combat, and competition among work groups, through which people bond into teams by learning to trust one another."[10] Trust and love encompass more than drives, desires, and the working of our bodies. They instead represent holistic phenomena, comprising both physical and spiritual experiences. The eminent British biologist John Maynard Smith harbors similar sentiments. He can't understand why organisms have feelings at all. Biochemistry alone determines behavior, according to his reading of science, and the accompanying sensations of love, fear, happiness, and so on play no role. They emerge from the material world but have no effect on it. When pressed, Maynard Smith suggests that, though subjective experience has no scientific explanation, still it may have a metaphysical explanation. Its importance may lie not in the material world but somewhere beyond or behind the perceivable universe described by science.

Such sentiments needn't clash with the biochemical basis of love. Scientific researchers exercise caution when talking about the link between biochemicals and love. Insel and Sue Carter emphasize the complexity of the human brain, insisting that a number of other factors mediate any effect of oxytocin or vasopressin. "It is important to recognize that oxytocin is only one link in a very complex neurochemical chain necessary for maternal behavior," cautions Insel.[11] "Oxytocin and vasopressin provide interesting case studies for investigating the neural substrates of attachment," he writes elsewhere, "but neither . . . should be thought of as . . . preeminent for these behaviors."[12]

Experience may play a role, for instance. Zuo-Xin Wang and Insel document an experiment in which pups from the polygynous meadow vole were fostered to monogamous prairie vole parents. The mature fostered meadow voles exhibited increased parental care in comparison to their non-fostered cousins. This "suggests the importance of early experience for acquiring patterns of parental care," argue Wang and Insel.[13] For humans, statistics suggest that the abused have a high chance of turning into abusers, while children of divorced parents are likely to go through a divorce themselves. Experience plays a role in human behavior too.

The neurobiology of love does not imply rigid biological determinism. Room still remains for past experience, free will, and subjectivity.

LET'S-HAVE-'EM-BOTH

Other critics might acknowledge the validity of the biochemical basis of love yet retain hold over the more subjective elements of our experience. Such respondents set off to a good start, but instead of tackling the issue head on by attempting to unite the physical and spiritual, they often invoke a dichotomy, insisting that both elements coexist side by side but apart.

These "let's-have-'em-both" adherents could find extra ammunition in the case of traditions that treat the Divine's essential nature as love. "All right," they might say, "we accept that when we display love, biochemicals underlie our behavior. But God's love? God is a spiritual being, not a complex of chemicals. God *is* love, some other kind of nonphysical love." This sort of response hits the nail on the head, of course. It doesn't save the idea of spiritual love, however. How can two such different notions of love exist? Why have we always treated divine and human love as similar? Can we reconcile spiritual and biochemical love? Suggesting—as such respondents might do—that God evolved us to have similar qualities to him invokes divine intervention into evolution with its inherently random mechanism of genetic variation. Answering questions and objections such as these requires that we radically rework our conception of the Divine and of the Divine's relationship with us and the universe.

Such critics may also spy a second loophole. "What about *agape*, the Platonic love that we should show toward our fellow beings?" they might ask. "None of your scientific research indicates that love extending outside the family circle stems from oxytocin or vasopressin." True, but evolutionary theory does suggest that altruism springs from our genes. Our genes push us to behave altruistically. By helping others, we ensure an easier ride for ourselves, thus giving those same genes a better chance to propagate themselves via our offspring. The next chapter covers this theory.

CONCLUSION

Whichever way we look at it, we can't get away from our biology. Then the biochemical account must clash with traditional accounts of the spiritual nature of love. We're left with an unloving conflagration. A way forward does exist—a way that recognizes the value of both accounts and that faces the problems head on. It means that we must ask difficult questions. Is love a choice or an involuntary action? How does divine love relate to human love? The task of reconstructing our notion of the Divine and of the Divine's loving relationship toward us is daunting. These endeavors, though, will enrich our knowledge and coax science and spiritual thought toward a new era of mutual respect and understanding. For such gains, a challenge and a rough ride seem a small price to pay.

CHAPTER 5

Evolutionary Psychology: A Science Background

> Love is only the dirty trick played on us to achieve continuation of the species.[1]

The past few chapters and couple of interludes have presented a picture of disciplines and methodologies at odds with one another, producing different results and proclaiming opposing ideals. These disputes surrounding love and happiness don't stand alone. They represent specific examples of a larger debate, that between reductionism and holism.

REDUCTIONISM

Reductionism has played a role throughout the history of intellectual inquiry and occurs today in most quarters and in many guises. Consciousness offers one of the most popular stages for reductionism. Back in the seventeenth century, René Descartes insisted that we comprise more than bodies and bundles of brain processes; we also comprise a spirit or soul, he thought, the thinking part of our being that makes us who we are. Today, physicalism or materialism replaces Descartes' dualism. It says that we can explain the thinking being, the conscious self in terms of (reduce it to) the material brain together with its physical processes. Away goes any special status for thoughts, feelings, and inner sensations; they become nothing but quirky manifestations of the brain's machinery.

Reductionism, in a more general sense, also emerges in the objectivism–subjectivism debate. Do we see what is "really" in the world when we look at it? Would the world appear the same whether we look or not? "Of course," claims the objectivist. "Our perceptions reflect reality." "But hang on a minute," interjects the subjectivist, "the world as we perceive it is a human construction. We bring something of ourselves, our culture, our inheritance to the world each and every time we sneak a look." The objectivist embraces reductionism: the world is as it is, period. There's nothing more to it. We humans make no contribution. Not far behind this kind of thinking comes the idea that the social sciences and humanities add nothing to the supreme explanatory power of physical science and so become worthless. The "hard" sciences provide us with the only true insight into the workings of ourselves and our environment.

EVOLUTIONARY PSYCHOLOGY

The skirmishes over love and happiness represent modern-day examples of a debate that has raged between scholars for the past couple of decades over evolutionary psychology (previously called sociobiology). The scientific discipline of evolutionary psychology examines the biological basis of all forms of social behavior in both humans and animals. It rests on evolutionary theory, as does contemporary biology, and focuses on natural selection of those traits most conducive to survival and so to reproductive success. Just as evolutionary theory explains why we developed hands with four fingers and an opposable thumb (to use tools allowing us to manipulate our environment for our own ends), so, according to evolutionary psychologists, evolutionary theory can explain why we behave in the social environment as we do.

Spiritual thought, sociology, and psychology have traditionally provided explanations for human social traits like altruism, cooperation, and aggression by drawing on the ideas of inherited culture, social praise and censure, and taboo. Human evolutionary psychology now seeks to explain these kinds of traits in terms of genetic fitness or reproductive success, measured by the passing on of genes from one generation to the next. Since evolutionary psychology concentrates on our genes, apparently ignoring the individual person and social institutions, many people declare it reductionist. Not only does it overlook the explanatory power of our social and cultural infrastructure, but it robs us of our ethics and our spirituality—the very things that make us human.

EPIGENETIC RULES

How do the explanations of evolutionary psychology work? How can our social behavior—our morality even—translate into talk about survival and transmission of genes? At the heart of the explanations of evolutionary psychology lie "epigenetic rules," innate dispositions that guide our actions and our behavior. Evolutionary psychologists isolate two kinds of epigenetic rule. *Primary epigenetic rules* guide us in processing raw emotional and sense data. On walking into a room, we register in a split second that it is full of people, that loud music is playing, that the atmosphere is smoky. Primary epigenetic rules enable us to produce a complete picture from the fragmentary input of our five senses. *Secondary epigenetic rules* assemble inner mental processes such as conscious and deliberate decision making and the placing of values and morals. They kick us into action once we've registered a situation via primary epigenetic rules. A woman who stumbles into a room where her husband is forcing their young daughter to have sex may straightaway report the situation to the police. She may begin to show her grief and anger once the shock subsides. Her immediate action springs from the belief that her husband's behavior is deeply wrong and requires punishment; she deliberately and consciously informs the police because of the negative value she places on incest. Her secondary epigenetic rules guide this reaction.

Epigenetic rules form a part of our inheritance. Our genes have encoded them and passed them down from generation to generation. Why? Because, in the words of Michael Ruse, they "have proven their worth in the past struggles of proto-humans."[2] They have insured—and continue to insure—our survival. The innate high-speed ability to classify and make sense of a potentially dangerous environment pays off. We quickly move out of the path of an oncoming car and live. Our feelings of repulsion toward incest help prevent the weakness and poor health that repeated interbreeding produces. Epigenetic rules enable us to handle our environment as best we can, so maximizing reproductive success.

ALTRUISM

Altruism provides one of the most striking areas in which epigenetic rules affect human behavior. We tend to assume that people perform generous acts and help others to the detriment of themselves because they *want* to do this, because they consider it the right thing to do. Spiritual and moral codes bolster this belief. Evolutionary psychology tells a different story.

Biological "altruism" exists at three different levels, according to evolutionary psychology:

- We all know how much most parents will sacrifice for the sake of their children: they feed, clothe, and provide for their offspring, even if this means going hungry themselves. This may ultimately extend to sacrificing life; think how often we hear stories of parents heroically returning to burning buildings to throw their children to safety. Evolutionary psychology explains this kind of "altruism" in terms of percentages of genes. Each parent passes 50 percent of their genes on to each of their children, and so, by performing actions that insure our child's survival, we ensure the passage (and survival) of our genes.
- "Altruism" also extends to the wider family. People feel more inclined to come to the aid of family members than to help a complete stranger. We might feel comfortable ignoring an unknown beggar on the street, but our priorities will change if that same beggar is also our brother or sister. Genes once again explain this kind of altruism (which evolutionary psychologists call "kin selection"). By coming to the aid of one of our relations (someone with whom we share genes), we again maximize the chances of (some of) our genes passing along to the next generation.
- What about the situation where we help out people other than family, perhaps offering to babysit for friends so they can shrug off their parental responsibilities just long enough to enjoy a quiet romantic dinner together? How can this possibly ensure the transmission of *our* genes? Evolutionary psychologists again have an answer. They call this kind of altruism "reciprocal altruism": we help out people unrelated to ourselves in the hope that they will one day return the favor. By babysitting for us one day in the future, our friends make our lives easier, so helping us pass on our

genes. Our original act of generosity becomes worth the effort.

Ruse and a father of evolutionary psychology, Edward Wilson, describe the role epigenetic rules play in any of these outpourings of "altruism":

> We need to be altruistic. Thus, we have rules inclining us to such courses of behavior. . . . We need something to spur us against our usual selfish dispositions. Nature, therefore, has made us (via the rules) believe in a disinterested moral code, according to which we ought to help our fellows.[3]

"Altruism" plays a valuable role in the transmission of our genes—too valuable to remain in our hands as a mere whim—and so our genes have encoded a secondary epigenetic rule that guides us toward our belief in the moral goodness of altruism. Our biology—our selfish genes—roots our ethics.

Our biology goes further than that: our genes encourage us to behave in many different ways, all of which lead to reproductive success. Love and happiness stand out as two examples. Happiness, as we saw in chapter 2, can keep us on our reproductive toes. Though we never stay happy for long (our levels of expectation readjust), still we continue to strive for that elusive goal and so remain productive, stretching and improving ourselves, strengthening our genes and passing them on. The evolutionary psychological role of love stands out even more plainly. Our feelings of love facilitate the passage of genes more than does anything else. Love encourages us to seek out partners, to have sex with them, and to pass on our genes. It also encourages us to care for our offspring, maximizing their chances of survival, and so ensure that those genes continue down the generations. Our genes ground many of our motivations, aspirations, and values.

CHAPTER 5

THE STRONG LOVE COMMAND

Several spiritual traditions, however, preach that we should practice love toward all and sundry, not just toward our immediate family. Neither should we love in expectation of reward or return. As we saw in chapter 3, Jesus commands us to love our neighbors—even our enemies—as we love ourselves; Mohists believe we should follow the example of the Will of Heaven by loving everyone equally; and Hindus exhibit true devotion to Krsna by (among other things) exhibiting amiability toward all beings. Ruse calls Christianity's version of this the "strong love command": "to love everyone . . . family, friend, nodding acquaintance, and enemy" and to forgive our enemies "virtually without limit."[4] This all-giving, no-returning kind of altruism, technically known as "transkin altruism," presents a challenge for evolutionary psychologists because it seems to fly in the face of talk about selfish, self-perpetuating genes. The farther away people are from our immediate circle of friends and family, evolutionary theory says the less we should feel morally responsible for them. "Altruism" should, biologically, become less and less workable. Spiritual thought and genetics can agree on some types of altruism, but they can't agree on extreme forms.

Some evolutionary psychologists, such as Richard Alexander and William Irons, propose that transkin altruism evolved because it allowed our early ancestors to limit conflict within groups.[5] As a result of this, they formed larger, more coherent groups, becoming stronger and focusing their energies more productively. The question remains, however, whether such theories really capture the meaning of universal love as many spiritual traditions advocate.

Ruse thinks not. "Morality is a biological adaptation,"[6] he believes, and evolutionary psychology and (Christian) morality are incompatible. Not only does the strong love command run counter to evolutionary expectations, but, Ruse insists, it even more importantly runs counter to survival, one of the most fundamental human values. When someone abuses or harms us, our biological urges encourage us to retaliate; love and forgiveness fall to the bottom of our priorities. It makes no sense, from the biological point of view, to follow a moral dictate such as the strong love command. Ruse further emphasizes that "most people would think it quite irresponsible to let someone else sin against them [innumerable] times." He feels "uncomfortable with a god who demands" this behavior of us, something we would naturally consider "morally perverse."[7] Most of us would agree with its perversity. We feel no sympathy toward child abusers or people who harm their spouses. Neither do we expect their victims to show sympathy toward them.

Ruse drives his final nail into the coffin by maintaining that biology destroys the traditional spiritual notion of morality. Believers "surely" place their faith in "an independent, objective, moral code—a code which, ultimately, is unchanging, and not dependent on the contingencies of human nature." Biology can't support such a moral code. Thus, "morality is just an aid to survival and reproduction, and has no being beyond or without this."[8] For Ruse, atheism remains the only option.

The Ruseian position sounds convincing; his arguments come across loud and clear. His standpoint embraces the reductionism that many people take as characterizing evolutionary psychology because he subjugates everything to biology, leaving no room for culture, society, or spirituality. Yet does evolutionary psychology necessarily entail a reductionist approach? Must we simply oust our cultural and spiritual heritage and have done with it? Reductionism may represent the Achilles' heel of Ruse's position.

CHAPTER 5

GENE–CULTURE COEVOLUTION

So far, we've talked only about the role of genes in reproductive success. We arrive in this world with epigenetic rules intact, encoded in our genes, but not only did our genes do this work. Our culture also played a part. Genetic evolution takes place in a social and cultural environment. Culture adds to what the genes generate in these dispositions by affecting the kinds of dispositions encoded. Culture changes genes. Evolutionary psychologists call this process "gene–culture coevolution."

Marc Feldman of Stanford University points to milk drinking as an example of gene–culture coevolution. To digest cows' milk, we must produce the enzyme lactase (which facilitates the absorption of lactose, milk sugar); if we can't, we become sick. Up to 90 percent of people in societies that have drunk milk for around three hundred generations can produce lactase. Four out of five people in societies that don't have this history, however, carry a different version of the enzyme and so become sick when they drink milk. Successful absorption of lactase by people in a culture where milk drinking is rife becomes important for survival and the passage of genes, and so our genes adapt to the cultural conditions. Genes and culture act in tandem.

THE SPIRITUAL ROLE

The gene–culture process might work in a similar way for more complex adaptive traits like transkin altruism. Donald Campbell believes that, when we consider human behavioral dispositions, we should pay attention to "culturally inherited baggage"[9] that we acquire through example and indoctrination as well as to the biological. "I am convinced that in past human history," he adds, "an adaptive social evolution of organizational principles, moral norms, and transcendent belief systems took place."[10]

Campbell believes that human beings living in contemporary urban civilizations practice self-sacrificial (or transkin) altruism. Common examples of honesty (returning a lost wallet to its owner) and generosity (donating money to charity) convince him of it. Yet biological evolution can't produce this type of altruism because it places potential cooperators in genetic competition with one another; our genes look blindly toward their own transmission regardless of the cost to our fellow beings. Social evolution has therefore had to counter biological evolution to allow for the dual development of transkin altruism and complex social organization. We therefore feel prompting from behavioral dispositions that incline us toward valuing society over and above ourselves, if social systems like ours are to work effectively. So enters spiritual thinking. "Committing oneself to living for a transcendent God's purposes, not one's own, is a commitment to optimize the social system rather than the individual system," explains Campbell.[11] The better a society's spiritual tradition functions, the more the society flourishes, the better life becomes for its members.

Scientist and spiritual thinker Ralph Burhoe also believes that biological "altruism" can't adequately explain transkin altruism. So, extending Campbell's work, he suggests that we comprise not only genetic but also sociocultural organisms, "information packets" that contain social constructions like language and spiritual thinking. (Oxford University zoologist Richard Dawkins terms such information packets "memes," giving, as examples, "tunes, ideas [including the idea of God], catch-phrases, clothes fashions, ways of making pots or of building arches."[12]) These "culture types" influence our behavior similarly to how genes do, subtly making us take on the language and spiritual thinking of the sociocultural context and heritage in which we participate. Further, natural selection works on culture types as well as on genes, selecting them independently of but alongside each other, creating sociocultural goals and genetic goals that coexist in harmony. Thus, adaptations emerge that benefit the organism as a whole both genetically and culturally.

Spiritual tradition, in Burhoe's scheme, represents "the agency which remembers and culturally transmits the basic long-range values or goals of the sociocultural organism."[13] Spiritual traditions act as the chief storehouses and propagators of our culture's most basic values. Via epigenetic rules, they instill these values and goals within us. They ensure that we live and move through a sociocultural "habitat," a "living envelope" holding information that both "protects" us and guides us "into a larger life."[14] Our brains, having received the spiritual traditions' transmissions, reflect a proportion of this culture type, particularly the basic values and goals. We feel responsibility toward our culture and society and toward our physical bodies since we comprise both culture types and genes.

Thus, transkin altruism comes into being. Our inherited genes selfishly demand survival. Our inherited values teach us to respect not only ourselves and our right to survive but also others and their right to survive. Genes and culture evolve in tandem to produce transkin altruism. By serving society, we also serve ourselves, which leads to growth and flourishing for all.

CONCLUSION

Perhaps evolutionary psychology lacks the reductionism many believe it entails. After all, at least some of its proponents allow a role for culture, society, and tradition; evolution doesn't take place solely at the level of genes. The interesting question becomes the extent of the role culture plays. Can culture break away from biology completely, or will our genes always hold ultimate sway over society? Ruse insists that ethics "is an illusion fobbed off on us by our genes,"[15] assuring us that atheism remains the only sensible road. Must we reduce morality and spirituality to nothing but a beguiling genetic conjuring trick, or can we reconstruct ethics and spirituality in the light of the teaching of evolutionary psychology? The success of evolutionary psychology compels us to answer these questions.

CHAPTER 6

Evolutionary Psychology: A Spiritual Background

> Ordinary morality is more creative than they seem to recognize in their theory; and probably their own morality is more ambiguous, self-deceptive and manipulative than it seems to be to them.[1]

Evolutionary psychology poses a threat to spiritual thinking and beliefs, as the previous chapter showed. Morality apparently represents nothing more than a sophisticated vehicle for genetic survival. True altruism vanishes, and we become mean gene machines. Spiritual thinkers perceive the threat and respond. Many add a twist to their battle with evolutionary psychology, claiming that morality has nothing to do with biology. So they introduce the familiar dichotomy (or dualism) between the objectivity of science and the subjectivity of the spiritual, assigning the spiritual to the sphere of value and relegating science to the realm of facts. Their response is as destructive as the original attack.

THE NATURALISTIC FALLACY

Philosopher Gerrit Manenschijn captures one of the most pervasive theoretical objections to evolutionary psychology with the following words:

> The prescription to go from "can" to "ought" can only be justified by the method of moral reasoning. Here scientific reasoning has no task. Of course, scientific reasoning can explain why some person in given circumstances in fact goes from "can" to "ought," but only moral reasoning is competent to justify the prescription that somebody ought to do what [they are] able to.[2]

This is the "naturalistic fallacy," or the distinction between "is" and "ought." We commit the naturalistic fallacy when we deduce how we *ought* to behave from how we *do* behave, when we make the illegitimate move from is to ought. Sixty-four percent of men have had sex with someone else while in a relationship, proclaims a recent *Arena* survey, yet according to thinkers like Manenschijn, this fact should not lead us to the conclusion that we *ought* to conduct affairs outside of our permanent relationships.

The naturalistic fallacy originated in the work of the eighteenth-century philosopher David Hume, resurfacing with Cambridge University philosopher G. E. Moore. Myriads of philosophers and spiritual thinkers now employ the fallacy to justify the continued separation of science and spiritual thought, biology and ethics, fact and value.

Evolutionary psychology commits the naturalistic fallacy, many spiritual thinkers believe. Evolutionary psychology as a science can only describe how we do behave and must make no claims about how we ought to behave. Michael Ruse's and Edward Wilson's suggestion that morality and altruism represent biological adaptations fails on this count. Biology and morality may coexist, but logically, they don't and can't inform or influence one another.

Brandishing the naturalistic fallacy in defense of spiritual thought bears a strong resemblance to a magician pulling a rabbit out of a hat. As with most magic tricks, things are not all they seem.

One common objection against the move from "is" to "ought" involves claiming that something natural does not necessarily qualify as something good. Evolutionary psychology doesn't suggest, though, that our biological drives and desires count as good. Evolution has armed us with conflicting motivations (should I stay in and care for the kids, or should I go out for a romantic night on the town leaving the kids home alone?), and it remains up to us to sort out the good from the bad. Further, all human moralities share some common elements. Cross-cultural studies show, for instance, that "altruism" and "reciprocal altruism" form the basis of all moralities. They represent the good for all peoples, past and present, and should continue to do so regardless of the various embellishments and details introduced by different societies. In such cases, the "is" can become an "ought" and probably should do so for the effective functioning of the moralities.

The soundness of the reasoning behind the naturalistic fallacy requires questioning. We employ the science of nutrition to determine what types of diet most benefit us; the "is" of science here determines what we ought to eat. We consider this move legitimate, yet when evolutionary psychologists suggest that biology helps develop morality—helps determine how we ought to behave—all hell breaks loose. Is the furor justified, or are we simply overprotective of our particular values? Overprotectiveness can produce damaging results. By erecting a barrier between the "is" and the "ought," between science and morality, we insulate the one from the other. This doesn't reflect reality. Values always guide science—for instance, in terms of direction of research and application of results—and values always revolve around how we do and do not behave.

REDUCTIONISM AND DETERMINISM

The claim that the "is" has no role in determining the "ought" in the context of evolutionary psychology encompasses the belief that genes have no control over culture and don't contribute to it. We fear reducing our culture (and so our spirituality and morality) to mere biology. Biochemist and Anglican priest Arthur Peacocke shares this fear. While he admits that evolutionary psychology has greatly enhanced our understanding of ourselves, our psychology and motivations, he still believes that it can't tell us much about the significant elements of human behavior. Spiritual thought still retains a powerful explanatory role:

> The scientifically reductionist account [of evolutionary psychology] has a limited range and needs to be incorporated into a larger theistic framework that has been constructed in response to questions of the kind, Why is there anything at all? and What kind of universe must it be if insentient matter can evolve naturally into self-conscious, thinking persons? and What is the meaning of personal life in such a cosmos?[3]

Only spiritual thought can provide adequate answers to such questions, and though the physical and biological worlds ground our mental and spiritual aspirations, only spiritual thought can recommend what we *should* aspire to.

Peacocke's primary concern centers on the reductionism he considers inherent in evolutionary psychology. Evolutionary psychologists tend to generate "confidently and explicitly deterministic, reductionist and functionalist"[4] accounts of human behavior. Their approach, he adds, rests on the assumption that we can isolate a specific gene for each type of behavior, even though we have no knowledge of the actual causal chains linking genes to behavior. He thinks that this renders the application of such an account to human behavior controversial at least.

Peacocke wrote these objections several years ago, and since then the science of behavioral genetics has come a long way. Research now shows, as we saw in previous chapters, that genes do underpin a number of our behavioral traits. We can expect this trend to increase as more studies emerge. The scientific understanding of the connection between genes and behavior also becomes ever clearer, though specific causal chains may still remain obscure. Neither do scientists any longer cling to the single gene hypothesis; more recent research teaches that large numbers of genes may influence a single behavioral trait. In the words of Dean Hamer, "Almost all behavioral traits are almost certainly influenced by many, many different genes acting together with one another and the environment."[5] New genetic research quells Peacocke's objections, and we can expect it to continue to do so even more effectively.

Along with Peacocke's fear of reductionism comes his distrust of those evolutionary psychologists who, he considers, propose that genes determine our social behavior. He admits that research may confirm evolutionary psychology, but he can't accept that genetics will explain all of culture. Do evolutionary psychologists really show that genes *determine* social behavior? Culture plays some kind of role, they allow, but still they hold that the cultural ultimately connects with the biological. Culture can never achieve total independence from biology. Wilson captures the contention well: "Can the cultural evolution of higher ethical values gain a direction and momentum of its own and completely replace genetic evolution? I think not. The genes hold culture on a leash."[6]

Wilson's talk of leashes and the tethering that this implies arises in the work of Donald Campbell and Ralph Burhoe, as the previous chapter discussed. Cultures as well as species evolve, they suggest, and spiritual thought prompts us to feel responsible for the survival of both our culture and our genes. The leash from biology to culture (including to spiritual thought and morality) stretches a long way on this account but not far enough for Peacocke. He questions the assumption—implicit in such accounts—that *survival* (of culture, genes, or anything else) counts as a *value*. "Survival for what?" he asks. "Is survival a value? What kind of survival?"[7] He wants rather to look at the *truth* of spiritual claims and beliefs before considering whether or how they contribute to our survival. Their truth may even suggest the irrelevance of survival for humanity and culture. Here Peacocke reminds us of Job's fearless pronouncement: "If [God] would slay me, I should not hesitate."[8] Value and morality remain, in Peacocke's eyes, the province of spiritual thinking, not biology. The leash doesn't exist.

Peacocke responds to Ruse's challenge that ethics is "an illusion fobbed off on us by our genes to get us to cooperate" with equal energy. While natural selection has, in the long-distant past, favored genes predisposing us to act "altruistically," it has also, Peacocke asserts, favored genes giving us "the capacity to think rationally" and "to evolve belief systems of our own devising."[9] He thereby draws a distinction between the *origins* and the *content* of morality, insisting that biology can account for the former but not the latter. Thought and knowledge, not drives and desires, dictate the particulars of our moral systems. "Ethical codes . . . are, in any case, not uniform, and are often counter-biological,"[10] he adds with a final twist of the knife. The tale of the Good Samaritan illustrates how highly developed moral codes run counter to biology (selflessly helping *any* person in need regardless of genetic relationships), so ousting more "primitive" moralities based on survival of individuals, communities, or genes. Evolutionary psychologists therefore stand guilty of the "genetic fallacy" in which the cultural or biological origins of culture can completely explain its development. "Just as science is not magic, so ethics, on the same grounds, is not genetics."[11]

Peacocke might sound convincing, but the points of disagreement between his position and evolutionary psychology amount to less than he appears to suppose:

- Evolutionary psychology doesn't say that we are the blind servants of our genes. Evolutionary psychologists would agree that we've evolved the capacity to think rationally, which allows us to devise our own moral systems. In fact, gene–culture coevolution relies on this scenario. Culture constructs a society's moral system based on biological requirements. Culture determines the good and the bad by comparing and contrasting various epigenetic rules, then combines and develops them to form a coherent morality. Culture (and so the individuals making up that society and culture) can and does influence and alter biological norms.
- Biological drives can conflict, and we must resolve these conflicts by rationally assigning positive and negative values. In a famine situation, we might experience inner conflict: should we satisfy the drive to feed ourselves, or should we satisfy the (equally strong) drive to feed our children? Those who neglect their children in favor of themselves attract censure; those who sacrifice themselves for their children attract praise. Culture assigns a negative value to the "me first" impulse and a positive value to the "my child first" impulse.
- Sometimes culture assigns values that override biology altogether. Most societies today aspire to monogamy as an ideal, though this opposes the biological drive to spread our genes far and wide.
- Evolutionary psychology also allows for discrepancy between ethical codes. Morality represents a fluid dance between biology, culture, and environment. Environments and cultures differ, and we adapt to circumstances in alternative and changing ways. A president's

> ability to run the country has little to do with that president's private life, many people in the United States believe. Yet fifty years ago, evidence of personal impropriety may well have provoked public outrage. Culture, morality, and motivation, according to evolutionary psychology, are always on the move.

Peacocke's real bone of contention with evolutionary psychology concerns the extent of culture's input to morality. He wants culture to display genuinely emergent features that distinguish it from all things biological. He wants to cordon culture off from biology, placing culture above the barrier and biology below it. Spiritual thinking and morality thus escape the taint of baser desires, remaining free to minister to our higher consciousness. To prescribe this separation can drive a dualistic wedge between science and the spiritual, evolutionary psychology and morality, fact and value, origin and content. Science and the spiritual in this view can speak different languages, represent different frameworks, and operate on different levels. Peacocke rightfully removes the dogma of reductionism, only to replace it with the equally damaging possibility of dualism.

Evolutionary psychologists, by contrast, don't succumb to the cordoning-off approach. They believe, to return to Wilson's metaphor, that genes tether culture no matter how much it might appear that culture roams free. Should we, like Peacocke, describe this position as reductionist? No. Thinkers like Burhoe and Campbell allow our cultural inheritance to play at least as big a role as our genetic inheritance. Culture doesn't reduce to biology, though the two are linked and progress hand in hand. Gene–culture coevolution describes not a reliance of the cultural on the genetic but a symbiotic relationship between the two. Peacocke's position lacks this symbiosis. While his motivations remain vitally important (rejecting reductionism and upholding an explanatory role for spiritual thought), they can lead down the blind alley of separation and dualism rather than toward the open road of reconciliation and holism.

THE SOVEREIGN INDIVIDUAL

John Bowker of Gresham College in London also responds to the challenge of evolutionary psychology but adopts (in some ways at least) a different stance from Peacocke. He, like Peacocke, laments the "poverty" of evolutionary psychology. It fails to "give an adequate status to the possible validity of human insight and judgment" by attempting to explain everything about human behavior in terms of genes and the unfolding of genetic programs.[12] Evolutionary psychology robs us of our humanity, he fears, and in so doing denies that we can interact with a spiritual reality (or deity), a reality "which does not appear as an object among objects."[13] Like Peacocke, Bowker stresses that genetic explanations fail to account for the beauty, virtue, and truth of spiritual experience. Unlike Peacocke, Bowker doesn't assuage his fear by driving a wedge between genes and culture. Rather, he admits that "in view of the highly complex relation between genetic and cultural evolution," we find it impossible to determine "the balance between genetic determination and cultural constraint."[14] Human nature doesn't lend itself to a straightforward division into the cultural and the genetic.

Bowker enlarges on these thoughts in his book *Is God a Virus?* He argues here that evolutionary psychology and gene–culture coevolution encourage us to ask misleading questions like, "Do the genes construct the cake on which culture simply deposits the icing? Or do the genes supply the ingredients which are then made into very different kinds of cake by the recipes of different cultures?"[15] For Bowker, such questions remain meaningless; we can't separate cultural from genetic contributions, nor do we humans simply represent the sum total of our cultural and genetic inheritance. Rather, both genes and culture contribute to the developmental process that characterizes each unfolding human life.

Evolutionary psychologists interpret genes and culture as causes that force us into one mold or another, insists Bowker, and Wilson's notion that genes always tether culture leads us to assume that, ultimately, only genes exercise causal power over our behavior. Bowker instead claims that genes and culture contribute "in varying ways (very often causatively), to the formation of human life, character, and behavior,"[16] but since each person develops differently, the effects will vary from person to person. Genes and culture do no more than set limits on human developmental possibilities. For Bowker, then, the individual becomes sovereign. The human organism, he claims, takes "control over its own outcomes within the boundaries of constraint," and "it is precisely this which creates the freedom of human nature to transcend the biological and cultural points of its departure."[17]

Evolutionary psychology (especially in its more recent formulations) doesn't ignore the potential or creativity of the individual, whatever Bowker may think. It acknowledges that, of course, we can rise above our genetic and cultural conditioning. Genes and culture don't mold us. Instead, we manipulate our inheritance (both genetic and cultural); we make ourselves what and who we are. "One of the biggest myths is that something is genetic therefore it is fixed. And of course this simply isn't true," explains Dean Hamer. "All these genes do is give us a disposition one way or another. Whether we act on that is still very much a matter of free will or choice."[18] He draws the distinction between temperament and character, explaining that "temperament is what you are born with. Character is what you've learned."[19] For instance, some of us arrive in this world with a genetic predilection for thrill seeking. Some fear nothing and will do anything: hang gliding, parachuting, bungee jumping. No one can predict what thrill seekers will make of their predilection because that remains up to the individuals concerned. A thrill seeker may become a firefighter or a drug addict; we may manipulate our genes for good or bad. This transformation of temperament into character remains our work and ours alone.

Bowker speaks of us "transcending" our "biological and cultural points of departure," yet he fails to specify the form that this transcendence takes. We shouldn't be surprised, perhaps. We can't (as Bowker rightly points out) separate genes from culture, but neither can we separate individuals from their genetic and cultural influences. Both culture and genes ground individual responses and behavior. Natural thrill seekers might create themselves either as firefighters or as drug addicts, but we must recognize that neither creation depends on free will alone. The choice between firefighter and drug addict represents a product of our Western, twentieth-century, largely urban culture. Alternative cultures exist (and have existed) in which people can't readily obtain drugs and where a coordinated fire service with which we're familiar doesn't exist. In such cultures, born thrill seekers can't choose between "drug addict or firefighter"; their sense of adventure must find its expression elsewhere, in the peculiar opportunities provided by their own cultures. Our expressions of free will, in this way, always connect intimately with both the genetic and the cultural landscapes. We transcend our biological and cultural points of departure but always within limits. We make ourselves what we are, but the "what we are" always forms part of a larger picture. Bowker, in his concern to stress the intertwining of genes and culture and the transcendence of the individual, fails to appreciate this issue.

CONCLUSION

We see the kind of response that evolutionary psychology draws from spiritual thinkers. We've looked only at Peacocke's and Bowker's reactions, yet their positions remain representative. Spiritual thinkers perceive their task as one of beating the enemy into submission. They scent a threat, and their response becomes extreme: evolutionary psychology screams reductionism, it disregards the individual, it belittles human rationality and creativity, it denigrates culture, morality, and spiritual thought. Yet none of these provide an honest picture of evolutionary psychology, which neither loses sight of the individual nor leads to an impoverishment of our cultural, rational, and creative achievements. Rather, it acknowledges that individuals, genes, and culture continuously work *together* to produce the fluid sociocultural environment that we inhabit and can subtly alter every day of our lives.

INTERLUDE 3

Philip Hefner's Conflagration Band-Aid

> While the mutual exclusion of scientific notions of nature and [spiritual] notions of ultimate nature has been common, does it need thus to continue?[1]

Many spiritual thinkers ignore the implications of evolutionary psychology for spiritual thought. Others attempt to meet the evolutionary psychological challenge but, as the previous chapter showed, decline a true reconciliation of science and spiritual thought. They often instead erect a dualism between the two by insisting that evolutionary psychology can't do justice to humanity, to our cultural, rational, and spiritual achievements. A small number accepts the evolutionary psychological challenge head on. This minority acknowledges the importance of evolutionary explanations and tries to present a spiritual picture that incorporates and works in tandem with scientific insights.

One such thinker is Philip Hefner, retired professor of systematic theology at the Lutheran School of Theology at

Chicago. He aims to "work with the methodological assumption that [spiritual] theories should be commensurate—though not necessarily identical—with scientific theories about the world."[2] This assumption leads him to reject several of the defensive maneuvers that the more conservative spiritual thinkers employ.

ANSWERING OBJECTIONS

THE NATURALISTIC FALLACY

Hefner concentrates on the history of evolution, on its seventeen-billion-year past and on what this tells us about our present and future. "The sciences are uncovering basic needs that are constitutive of the human being from its evolutionary past," he explains, "and they are suggesting that these needs comprise a system of values that is necessary for human life and for the life of the planetary ecosystem in the future."[3] The naturalistic fallacy represents a false dichotomy for Hefner. False because evolutionary psychology describes our wants and needs (the "is"), and in doing this, it defines an imperative; it "sets parameters that must be satisfied if we are to exist" (the "ought").[4] We need love and companionship—our well-being depends on it—and so we feel the (biological) urge (that culture enforces) to cherish and care for our children, and to love and support our partners. Our evolutionary past grounds how we are now (the "is" of today) and how we are today implies how we ought to behave, both today and tomorrow. "Is" and "ought" become two complementary sides of the same evolutionary coin.

SURVIVAL

"*Is* plus *ought* equals survival," insists Hefner.[5] Recall Arthur Peacocke's queries: "Survival for what? Is survival a value? What kind of value?" Hefner provides an answer. The Judeo-Christian tradition asserts that the Divine created the universe. Evolutionary theory teaches that survival via natural selection comprises an intrinsic component of that universe. Thus, concludes Hefner, "the framework of the survival process and the mechanisms of survival are part of what [the Divine] has created."[6]

Hefner doesn't doubt that survival represents a value. The Divine's initiation of natural selection gives survival top billing, and since many spiritual traditions teach that the Divine shows ultimate love and faithfulness toward creation, we must assume that survival (rather than destruction) figures high on the divine agenda, that survival counts as a value. Survival relates to the Divine's purposes toward the universe, adds Hefner. "This does not mean that God's purposes are subsumed under the mechanisms of genetic survival or species survival or even ecosystem survival." It simply means that "these mechanisms are part of something larger, namely the divine intentions."[7] Survival represents a value because it emerged from the Divine and forms part of the Divine's master plan.

LEASHING

What, then, of Edward Wilson's leashing question? Hefner emphasizes the intertwining and mutual dependence of biological and cultural evolution. Though biology and culture "flow in different channels," he explains, still "these channels merge in the human brain," jointly giving rise to our behavior, morality, and spirituality.[8] Our biological development enabled—or gave rise to—our cultural development, and so the two dimensions exhibit continuity and consistency. Culture's biological rootedness fails to perturb Hefner: "If altruism in culture really does oppose the genes, it will be signing culture's death warrant, since genetic evolution is the host in this symbiosis. Without the biological host, culture cannot enjoy the party."[9] Yet culture has its own method of payback, adds Hefner, because it can refuse to cooperate with biological urges. Pushing our way to the front of the line may ensure that we get served first, but it does nothing for our relationship with patient customers. Culture instills the "wait-your-turn" morality in us, regardless of how selfish or impatient we might feel, and for good reason: if everyone pushed to the front, no one would get served, and disorder would ensue. The best solution in this situation requires biological limitation.

A RECONSTRUCTION

That many spiritual thinkers find survival via natural selection difficult to understand or explain, that evolution seems to entail reductionism and a diminished role for spiritual thought—these all become irrelevant, according to Hefner. His answering some of the most common spiritual objections to evolutionary psychology opens the stage for reconstruction. Spiritual thought must turn around and face such difficulties, not bury its head in the sand. Hefner effects his reconciliation, first, by taking science seriously. "Since the scientific understanding of the world is one of the chief ways in which we encounter this world and know it, the task," he explains, "is to discern how the world scientifically perceived is referable to God and how it can be the *effect* of God."[10] Evolutionary theory, natural selection, and evolutionary psychology strike at the heart of spiritual thinking, but this should only strengthen our resolve to find a solution that does justice to the insights of both science and spiritual tradition because these sciences describe God at work.

Hefner finds spiritual thinkers' ignoring the role of natural selection "embarrassing . . . since all of life is woven on the loom of selection processes."[11] Evolution is responsible for the past, present, and future of the universe. It is also responsible for the past, present, and future of humanity, for our social behavior as well as for our physiology. The question becomes, How can we remain faithful to scientific teachings while retaining spiritual beliefs? Hefner's response portrays the Divine as initiating evolution and natural selection; these processes form part of the Divine's vision for the universe. The Divine stands as sole creator of the universe, and its sustained unfolding depends on divine purposes. Yet evolutionary processes remain real because they make the universe how and what it is. Hefner pushes the point further: "Is [this] cosmos essentially friend or foe?"[12] Friend, he believes. We needn't fear the apparently blind forces of natural selection since the all-loving, all-giving Divine has created and sustains those forces. The cosmos remains "a trustworthy environment for the human. . . . It is a fundamentally friendly home for us."[13]

We play a special role in the divine cosmos, according to Hefner's theory. Evolutionary forces not only determined our physical and social capabilities. They also handed us the gift of freedom: the gift of reflection, of deliberation, and of the ability to come to decisions and to act on those decisions. Along with all this comes the task of taking responsibility for our actions, of recognizing and righting our freely made mistakes.

Hefner reasons that this freedom falls within the compass of divine purposes because evolution gave freedom to us and the Divine purposefully sustains evolutionary processes. He follows this line of thought further:

> Such reflection has suggested to me that *Homo sapiens* might be considered to be, as its very essence, a *created co-creator*. "Created" in that *Homo sapiens* was defined prior to its actually assuming its role in the evolutionary process, created by God through the process of evolution itself. "Co-creator" in that God has in significant manner rendered the entire divine work of creation (at least on planet earth) vulnerable to the creating work of the human species.[14]

Our dual, determined-yet-free nature means that, while we remain the Divine's creations (via evolutionary processes), still we hold some sway over our own and our planet's destinies. We have the potential to alter the unfolding nature of the created universe. In this potential lies our destiny, the "future toward which we are being drawn by God's will," a future in which the Divine plans "to perfect all that constitutes the creation, human beings included."[15]

Hefner convincingly attempts to draw the scientific and spiritual camps closer together and to heal the wounds sustained by heavy bouts of sparring. He stands out as one of the few spiritual thinkers with the necessary conviction and motivation to achieve this. Saving him from falling into the dualist trap—potting, Hefner implies, its dry and barren road—is his belief that we should take science seriously, in particular that evolution governs species' survival. We can't meaningfully distinguish the "is" from the "ought"; survival remains perhaps the most important human value, and biological and cultural evolution exhibit fundamental codependence. More generally, science and spiritual thought don't speak different languages, exist in different realms, or evaluate different phenomena. The universe remains unitary, Hefner believes, and scientific and spiritual thinking must describe that universe in a complementary and holistic manner.

PROBLEMS WITH THE RECONSTRUCTION

DIVINE PURPOSE?

Does all lie well with Hefner's account, then? Have we found the solution to our questions? Unfortunately not. Hefner's solution assumes too much.

"What is the significance of the world for God?" This, according to Hefner, represents "the most basic [spiritual] question." Behind the Divine's actions, Hefner answers, lies divine purpose. The Divine intends to complete and perfect the created universe, to "consummate" it.[16] Natural selection, the development and survival of the human species, and the evolution of human freedom—all of these contribute toward the final, perfected product. These are tools by which the Divine is bringing about the Divine's intentions.

Why should we assume, however, that purpose drives the Divine at all? Purpose can certainly drive us humans. We work to provide for ourselves and our families, we plan holidays for the future, we decide on the right time to apply for promotion or a new job. To assume that the Divine also lives this way is to project our intrinsically human qualities, our own motivations and purposes, onto the Divine, and we have no justification for doing this. Must the world hold significance for the Divine? We certainly *hope* so. We want to feel cherished and cared for, we want to feel important, we want to feel that our life and achievements mean something on the grand scale. But these desires emanate from our human psychology, a psychology rooted in our biology and our cultural inheritance—a psychology, biology, and culture foreign to the Divine.

A HOSTILE WORLD?

Hefner assures us that, though evolution and natural selection continue apace, we needn't fear; we hold a privileged position at the center of the Divine's purposes. We have a divine destiny to fulfill, and our survival remains fundamental to that destiny. The universe thus constitutes "a fundamentally friendly home for us." How can Hefner be so sure? Evolution, natural selection, and genes are *blind*, science teaches. Only the fittest species survive. Right now, the human species remains successful, flexible, and fit, but this may not continue. New strains of virus keep evolving that remain immune to even the strongest antibiotics. The global climate is changing, and this will probably unleash horrific wars for the control of natural resources. We may overcome these difficulties, but on the other hand, we may not, and our fitness will suffer. Evolution may blindly turn from our friend into our enemy. A Divine who works within the confines of evolution can't suddenly jump outside natural processes to turn an unfavorable situation around. Hefner embraces the role of natural selection but not far enough.

Can the freedom that evolution bestowed on us save Hefner's attempted reconciliation of science and spirit? Hefner believes it deems us "created co-creators" whose job becomes "the modifying and enabling of existing systems of nature so that they can participate in God's purposes in the mode of freedom."[17] Yet freedom can lead to good or bad. The universe may currently represent a friend; we can, by our own actions, just as easily turn it into an enemy. Our disregard for the environment—our reckless use of nonrenewable resources, governments' refusal to halt destructive exploitation of resources, our reluctance to use public transport, lazy attitudes toward recycling—may facilitate this sooner than we think. Surely, Hefner doesn't have this kind of scenario in mind when he talks of our participating in the Divine's consummation of creation. Of course not. He insists that we must ask ourselves four questions when evaluating our actions: "Are we acting as if the planet is God's creation? Are we acting as if the planet were on its way to the consummation that God desires? Are we acting in full recognition of the role and purposes of *Homo sapiens* as self-defining co-creator, created as such by God? Do we acknowledge that the being of God is the final arbiter of our actions and their correctness or error?"[18] But these questions return us to our earlier objections: What justification do we have for assuming that the (spiritual) Divine shares with our (biological) selves the capacity to have purposes and desires? How can blind, opportunistic evolution contain the potential or direction for a perfect end product? It seems, despite Hefner's brave attempt, that we must evaluate and make decisions on our own after all. If the Divine doesn't share in our purposes, motivations, or desires, then morality (the "ought") comprises an essentially human (rather than divine) arena.

CONCLUSION

Hefner's conflagration Band-Aid doesn't cover the whole graze. Gaps still remain. His approach does surmount some tough obstacles, though: his refusal to accept dualism and his determination to meet and embrace science remain laudable. The attitude of reconciliation and cooperation that he shows moves us beyond the fragmentary positions sketched by Peacocke and John Bowker. But we need to run further with Hefner's reconciliation than he has.

The new sciences of behavioral genetics and neurochemistry throw up fresh challenges. Can the sciences of love and happiness provide new angles on the unification of science and spiritual thought? How might we more adequately project qualities onto the Divine? What might the Divine look like according to such a projection? How can a reconstructed spirituality offer its insights to science? These kinds of questions highlight the way toward reconciliation that the new generation of thinkers must follow.

INTERLUDE 4

Ted Peters' Conflagration Band-Aid

> The awareness of the Beyond is a matter of faith. Thinking about the Beyond is an intellectual activity, the structure of which we share with all other thinking activities. The Christian doctrine of creation as we have it, then, is a product of both revelation and reason, of both faith and science. It is the result of evangelical explication.[1]

Ted Peters, professor of systematic theology at Pacific Lutheran Theological Seminary, Berkeley, is a second spiritual thinker who recognizes the relevance of recent scientific discoveries for modern spiritual thinking. His account mirrors aspects of Philip Hefner's viewpoint, but, as we shall see, it also incorporates his own unique arguments regarding the reconciliation of science and spiritual thought.

HYPOTHETICAL CONSONANCE

Peters advocates what he calls "hypothetical consonance." He looks for "those areas where there is a correspondence between what can be said scientifically about the cosmos and what the [spiritual thinker] understands to be God's creation." Where we find consonance (accord or harmony) between science and spiritual thought, we should, he believes, further explore the possibility that they speak about the same thing. This position incorporates the assumption that there exists one Divine and one world, and so, "in the long run, science and [spiritual thought] are attempting to understand one and the same reality."[2] Science and spiritual thinking share points of contact, they speak about the same reality, they operate on the same level. Peters tries to avoid the dangers of dualism.

Modern thought, Peters contends, drives a wedge between the objectivity of the world as science describes it and the subjectivity of the way the human mind experiences it. But a new movement, postmodernism, is now emerging. Postmodernists seek to "reunite what has been separated, to fix what has been broken apart." They seek to reinstate "our naive sense of oneness with the whole of the world." Peters places himself in this movement, aiming to reunite science with spiritual thought and to explain the Divine's relationship with the world in terms of "future wholeness for all creation."[3] Hypothetical consonance, Peters's approach to science and spiritual thought, emerges from the unified holism of a postmodernism.

A RECONSTRUCTION

Peters reinstates our sense of oneness with the world much as does Hefner by concentrating on the Christian belief in the Divine's future consummation of all things, on the Divine's promise that, in the future, all things will become whole. We need to step back a little to view Peters's picture.

CREATION

The Divine, for Peters, is the creator of the universe, and this creation takes two forms. First, the Divine created the universe out of nothing. No material world existed prior to the Divine's bringing into existence the world as we know it. Second, the Divine now influences the world's continuing creation. The Divine underlies the ongoing progress of the universe from era to era, from primordial matter, through the dinosaurs' reign on earth, right up to the emergence and continued development of modern humans. Peters' position "depicts God as constantly engaged in drawing the world out of nonbeing and into existence with the aim of consummating this creative work in the future."[4]

Peters' interest in continuing creation leads him, like Hefner, to cut ties with the past and accord the future the pole position. Reality doesn't depend "on its past," he says. "It is cut free from the principles established at the point of origin. . . . The God of future salvation—the God beyond the present state of reality—is not restricted by what already exists."[5] Having created the universe from nothing, the Divine constantly guides creation toward its destiny, unconcerned by what has occurred in its past. The impetus always comes from in front, and so "God's present work in and for the world anticipates the final work."[6]

Creation, Peters believes, connects intimately with redemption and salvation in the following way. The (Christian) Divine forgives our sins and promises a final redemption for the dead. This promise indicates that the Divine has the power to transform the old order into something new. And "if God has this transformative power," Peters explains, "then God must have had the power to bring creation into existence in the first place. . . . The logic is this: the God who saves must also be the God who creates. Nothing less will do."[7]

THE BIG BANG, ENTROPY, EVOLUTION, AND ECOLOGY

All this describes a spiritual—not a scientific—world picture. How can Peters justify his spiritual beliefs in the context of scientific discovery? He kicks off by acknowledging the enormity of the task: "The most formidable challenge to [spiritual] thinking comes from the scientific community because modern science has produced staggeringly significant knowledge regarding the nature of nature."[8] He then examines four interrelated scientific ideas: the big bang, entropy, evolution, and ecology.

According to the big bang theory, all the material that now makes up our universe—everything that exists—originated as a single concentrated point of unimaginable heat and density within a reality undifferentiated by space or time. Roughly fifteen or twenty billion years ago, all that changed: the single point started to expand at enormous speed, and the cosmos as we know it today came into being as fragments or fallout flying through space away from the point of origin. This expansion will end, some scientists predict, when the explosion has dissipated all its energy.

This is where entropy (or the second law of thermodynamics) comes in. Entropy means that, within a closed system, heat flows in one direction only—from hot to cold—and that once all the heat dissipates, the system reaches equilibrium. No more heat energy remains to keep the system moving. (Other scientists predict alternative scenarios [contraction of the universe to a big crunch or the universe reaching a stable state where the inward pull of gravity equals the outward force of the initial explosive energy], but Peters ignores these.) If you leave a steaming mug of coffee on your desk, it eventually becomes cold, reaching equilibrium.

There's no hope of that coffee magically becoming hot again unless, of course, you introduce new energy with a zap in the microwave.

We can apply entropy to the cosmos as a whole. At some point in the future, all energy will dissipate, the big bang expansion will end, and the universe will die a cold, inert death, just like the abandoned mug of coffee. The cosmos has a finite life span. "For all practical purposes," explains Peters, "we can say time began when the big bang first began its bang."[9] He also believes that time will end at the point when the cosmos reaches complete equilibrium, when the universe dies.

Creation and development can still occur within a universe on its way toward equilibrium. Some parts of that universe will exhibit increased energy, leading to continuous fluctuations or changes. Planet earth constantly receives heat energy from the sun, and this energy allows many life forms (plants, animals, humans) to develop and flourish. "There are pockets or regions or subsystems that have increased energy," as Peters puts it, "and this combined with the interplay of randomness and chance provides the continuing creativity of the dynamic universe."[10] Change occurs at bifurcation points (or forks) produced by the continual fluctuation of these subsystems, but until change actually occurs, we can't predict what form that alteration might take. Imagine that your morning newspaper informs you that a new type of mouse has evolved: one that lives off burgers and pizza—the detritus of twenty-first-century life—and that avoids old-fashioned traps and poisons. Before its appearance, though, no one could have predicted its evolution. Some other junk food–loving species might have evolved, or, less plausibly, we might have changed our habits, disposing of scraps carefully, leaving our streets clean. We have no present window onto the future novelties of evolution.

Evolution and ecology thus arise. Planet earth, with its increased energy, undergoes creation and change, making possible here the gradual evolution of life. "Nothing was created all at once," Peters explains. "Everything, including human beings, is on the way. . . . We can expect still more creative activity in the future. Creation is ongoing."[11] Ecology (biological and physical conditions and relationships) maintains life on planet earth while allowing for evolution and facilitating ongoing creation. We humans form part of this ecology. We wield creative power and have the potential to influence the universe's forward march; our actions—throwing half-eaten burgers on the street, dumping toxic waste into the ocean, emitting greenhouse gases—affect the planet's future. Peters concurs with Hefner's notion of our role as created cocreators.

JUSTIFYING THE RECONSTRUCTION

Peters justifies his spiritual commitment to divine creativity in terms of these scientific ideas. The big bang and its associated notion of finite time mirror Christianity's belief in the Divine's creation out of nothing: time, space, and matter came into being at the moment of the big bang or of divine creation, depending on our point of view. "There may be some consonance between natural science and this Christian commitment," Peters insists.[12] The change inherent in thermodynamics and evolution also reflect the Divine's continuing creative work.

Yet Peters wants to go further than a simple match between scientific and spiritual ideas. He wants to justify his spiritual commitment to the future, to the Divine's redemptive promise of a new creation. He starts this with a hypothesis: "God creates from the future, not the past." The future, he believes, represents the only way forward. It sustains our present existence and our existence to come. "To be is to have a future. To lose one's future and to have only a past is to die. Deep down, we know this."[13] The Divine bestows the gift of the future on us, first by creating the universe out of nothing (producing the big bang) and then by facilitating change within the universe (via evolution), so allowing it to develop, to transform itself into something new. "The power of the future is the source of contingency or chance in nature and the source of freedom in human consciousness."[14] By opening up the future, Peters suggests, the Divine releases us from our past and allows us to deliberate, to act freely. We plan what to eat for Friday's dinner party, where to go on vacation next summer, and how to spend our retirement years, secure in the knowledge of tomorrow.

This creative activity of the Divine coexists with the universe's intrinsic development. It doesn't emanate from the outside, interfering in the natural course of events; rather, "every divine action within the world . . . is performed in, with, and under the things of the created world."[15] These two domains intertwine, interact, and nourish one another. Peters also believes that divine involvement with the world represents a purposive activity. He describes this as the "master act by which God creates and consummates the entire cosmos."[16] Consummation will reintroduce unity into a broken world. With this unity, Peters continues, comes our salvation: the Divine's ultimate purpose as the death and resurrection of Jesus Christ reveals to us.

PROBLEMS WITH THE RECONSTRUCTION

Peters paints a grand and unified picture, yet the richness of his account hides several flaws. Some of the criticisms applied to Hefner's account in the previous chapter also apply to Peters's position. Why should we think that purposes drive the Divine? Why must we expect any consonance between human and divine desires or actions? How can random cosmic and evolutionary processes purposively lead toward a consummation?

THE PULL OF THE FUTURE

Further criticisms pertain specifically to Peters's position. He emphasizes the significance of the future. The Divine's creative activity represents "a pull from the future rather than a push from the past"; the Divine always has future consummation in sight.[17] Yet it is unclear that we can interpret natural processes like the big bang, entropy, and evolution in terms of a pull from the future. Take evolution, for instance. Organisms evolve via natural selection. Those organisms and characteristics best adapted to their present conditions thrive and survive on into the future, while the least well adapted either evolve into something better or perish. Note the use of the word "present": natural selection acts on *present* organisms and *present* circumstances. Tigers have evolved lean, muscular bodies, razor sharp teeth, and protective stripes, all of which make them deadly killers, yet contemporary changes in their habitat (a marked decrease in numbers of prey, perhaps) may trigger the evolution of new characteristics as a survival response to change. The future doesn't influence evolution. How could it? We can't predict how the future will look because it remains at the beck and call of chance, as Peters rightly points out. Natural selection may determine which species and which characteristics will exist in the future, but natural selection doesn't work from the future. It constitutes a reaction to the here and now.

Perhaps Peters thinks that, though we can't envisage the future, the Divine can. This type of argument, however, brings us squarely back to the problem of purpose. Random natural processes and purposive consummation remain incompatible. Peters can't retain a belief in both, unless he also claims that the Divine can jump in and correct the universe when it runs off course. He thankfully jettisons this fanciful option.

SUBORDINATING SCIENCE TO SPIRITUAL THINKING

Peters appears to take science seriously, but his commitment runs skin deep. The big bang, entropy, evolution, and ecology represent points of contact (or consonance) between science and spiritual thought, he claims. We might agree. But instead of fruitfully exploring these points of possible connection, Peters succumbs to temptation, subordinating science to spiritual thought. He collapses the Divine's act of creation and the big bang, saying, "Not only does God release the exploding energy that drives the universe, but God also opens up the future so that new things can occur. The gift of the future is the very condition for the coming into existence and the sustaining of any present reality."[18] He believes that creation out of nothing and the big bang describe the same event. But do the two really equate? Creation out of nothing suggests, quite literally, producing something out of nothing, pulling a universe out of even less than thin air. Cosmologists agree, however, that something existed prior to the big bang. Something out there went bang. Their descriptions of this "something" differ, though not markedly. The big bang arose, suggests physicist Edward Tryon of Hunter College in New York, from events at the quantum level of "the vacuum," which comprises infinite energy and lacks air, space, and time. Other cosmologists aim for something more basic than the vacuum. Cambridge University physicist Stephen Hawking opts for an initial form of logic or mathematical consistency that produced the unique set of laws necessary for the evolution of our universe. John Wheeler of Princeton University introduces the "pregeometry": a primitive structure that underlies the "geometry" (shape or structure) that we usually attribute to our spatiotemporal universe.[19]

How do we reconcile these various notions of the pre–big bang "something" with the spiritual story? If the

Divine initiated the bang (or released the exploding energy, as Peters suggests), then the Divine didn't miraculously produce something out of nothing because something (the vacuum, an initial logic, the pregeometry) existed from which the bang banged. Does Peters need to modify his vision of the Divine? Perhaps the Divine permeates throughout reality, big banging right along with the universe and unfolding as the universe unfolds. Perhaps the Divine equates with the pre–big bang "something," comprising the conditions necessary for the evolution of our universe. Peters needs to examine these kinds of issues more closely. Both science and spiritual thinking offer lessons to teach us. By subordinating science to the spiritual, we allow their joint insights to pass us by.

FALSE GENETIC REDUCTIONISM

The new science of behavioral genetics also interests Peters because it offers further implications for spiritual thinking. He points out (as previous chapters also did) that we must be careful to distinguish between popular and scientific images of genetics. The popular notion of genes alone determining our behavior doesn't mesh with current scientific research programs, which assume that various factors (genes, the environment, and free will) together hone our behavioral traits. Yet he still takes issue with behavioral genetics over human freedom, doubting that our biology alone makes us free. He would rather invoke the Christian tradition in which "freedom is not merely a product of our biology but an achievement of divine grace working in the human will that takes us beyond our biology." It would be closer to the mark, he believes, "to say that freedom in the fullest sense stretches us beyond our given human nature."[20] Freedom becomes the domain of the Divine, not of our biology. Peters illustrates this with the now-familiar example of altruism. We can, he believes, rise above self-interested loving. We know how to love for the sake of love alone; we can, for example, love our neighbor as ourselves, love the Divine as the Divine loves us, even love our enemy. But none of this would occur, Peters insists, if genes alone governed our actions. Genes are inherently selfish. We need freedom if we are to overcome their selfishness. The Divine graciously provides this freedom, liberating us from the constraints of our biological natures and so allowing us to express altruism in the broadest sense. "Rather than being driven solely by the agenda of a human nature designed by selfish genes, we transcend our genetic code. Whatever human nature is, it is more than merely what we find in our genes."[21]

This account of genes versus freedom is, at best, naive. We should distinguish between popular and scientific images of genetics, Peters insists, but he then goes on to ignore his own advice. As the past few chapters have shown, behavioral genetics doesn't preclude freedom; we arrive in this world with certain predispositions, yet we freely choose how to interpret them and help mold our own personalities. We aren't mere puppets operated by genetic strings alone. Recall Dean Hamer's remark: "One of the biggest myths is that something is genetic, therefore it is fixed." Neither do biological accounts of human nature rule out broad-brush altruism. Room even exists for sympathizers of evolutionary psychology, such as Donald Campbell and Ralph Burhoe, to argue that our genes and our culture evolve together, producing complex adaptive traits like transkin altruism that benefit humanity as a whole. We needn't resort to the mysterious movements of the Divine to account for these kinds of phenomena. Peters once again succumbs to temptation and subordinates science to spiritual thought.

CONCLUSION

The emergency surgery provided by Peters fares less well than Hefner's conflagration Band-Aid. Though he recognizes the relevance of science for spiritual thought and so desists from driving a dualistic wedge between the two, the initial promise of Peters's position fails to produce. His heart lies with his spiritual beliefs, reducing science to spiritual thinking and insisting that the big bang, evolution, and human freedom depend for their origin and continued existence on divine processes rather than those of physics or biology. The oneness Peters claims to seek won't result from reducing one domain to the other, hoping to pass the product off as a unified whole.

Points of consonance between science and spiritual thought do exist, and we must press these points further if we want to arrive at a more complete understanding of the universe. To apply such pressure involves respecting the insights from each discipline and approaching the problem with an open mind. We may have to reconstruct our notions of the Divine and the Divine's relationship with the universe. We may have to review the status of scientific knowledge, so often interpreted in a deterministic, reductionist light. Recovery may prove painful, but the potential healing and fuller life make the procedure worthwhile. Peters misses a golden opportunity.

CHAPTER 7

Divine Projections

> We firmly believe and simply confess that there is only one true God, eternal, immense, unchangeable, incomprehensible, omnipotent, and ineffable.[1]

> God is a Spirit, infinite, eternal, and unchangeable, in his being, wisdom, power, holiness, justice, goodness, and truth.[2]

We long to understand, to gain knowledge, to describe. The science-driven twentieth and twenty-first centuries reflect this longing: we seek an intimacy with ourselves and our universe, and we seek to use our knowledge in multiple ways, through medicine, genetic engineering, and computer technologies. Similarly with the Divine. We yearn to appreciate the Divine and to understand the Divine's relationship with the universe and with ourselves as integral parts of that universe. We facilitate our understanding by using projections. In this activity, we turn toward familiar turf—human qualities and characteristics—and project them onto the Divine, turning the Divine into an exaggerated version of ourselves.

This tendency produces one of the primary problems in the conflagration concerning happiness and the one concerning love.

PROJECTIONS ONTO THE DIVINE

Projections typically produce an image of the Divine, within the Christian tradition at least, as a personal, loving, infinitely good patriarch who desires nothing but happiness for us humans ("his" creation). We view the Divine as the greatest conceivable being. Scholars mirror this predilection, ascribing awesome properties to the Divine. God is omnipresent (constantly present throughout the entire universe), creator and sustainer of the universe (as Ted Peters argues), perfectly free, omnipotent (all-powerful), omniscient (all-knowing), perfectly good (thus the ultimate arbiter of right and wrong), and everlasting (the Divine always has and always will exist). To conceive of the Divine at all, we conceive in terms we understand—power, knowledge, creativity, goodness—and then we magnify.

Similar tactics come into play with spiritual accounts of love and happiness. The Divine's essential nature comprises love, according to Christian tradition, and the Divine bestows that love on us openly and freely, as parents love and demand the best for their children. Christian Scripture urges us to take divine love as our inspiration, to return the Divine's love for us, to love our family, neighbors, and enemies as we love ourselves. Judaism thinks similarly, though with a parallel emphasis on justice and more importance placed on loving the Divine by following the divine Law. We should love everyone equally, Mohists believe, as does the will of Heaven. In that way lies peace and perfect order. The figure of the *bodhisattva* in Mahayana Buddhism expresses infinite love, shouldering our suffering and leading us toward the final peace of *Nirvana*, oblivious to any feelings of self-interest. Depictions of the Buddha with a thousand eyes and arms, all seeking out and assuaging our suffering, confirm our tendency toward anthropomorphism of divine figures. Islam encourages its devotees to follow the prophet Muhammad, to engage in the virtue of sexual love as a means of approaching the Divine. Recall Muhammad's words: "I fast, I break the fast, I pray, I sleep, I go unto women; beware! Whosoever deviates from my *sunnah* is not among my followers." Hinduism's Krsna represents a personal god; love for Krsna guarantees our salvation, and Krsna returns our love, graciously terminating the cycle of karmic rebirth, so bringing eternal peace. Divine love resembles yet surpasses our own. It stands as an example to us.

What about happiness from a spiritual point of view? The emphasis falls a little differently here. Perfect, unending happiness awaits us in the life to come, but (some traditions teach) we can experience partial happiness in the here and now. Contemporary orthodox Christianity focuses on the joy and reward of heaven, and modern spiritual thinkers believe in happiness arising from a serious, committed spiritual life. Judaism encourages us to do what's necessary to feel happy though not to carry what we do to excess and sometimes to give it up for higher causes. Virtue alone results in true happiness, according to Confucian thought. It encourages us to cultivate moral virtues and to share our successes with the wider social group. Taoists believe that we can't find ultimate happiness on earth. Rather, we must follow the Way, detaching ourselves from worldly pursuits, living in harmony with nature. Happiness awaits us in this "inaction." Buddhists understand ultimate happiness as annihilation of the illusion of the self, leading to the eternal peace of *Nirvana*, while Hinduism teaches that performing virtuous actions in this life leads to happiness in the next life within the cycle of rebirth.

We must, according to the Yoga system of Hindu thought, discriminate in meditation between nature and the self to end our suffering and attain happiness. Islam teaches that the Divine tests our preference for good and evil. Those who opt for good enter eternal *Al-janna*, experiencing the highest in sensual and spiritual joy.

Love and happiness coincide in some of these traditions. The Divine loves us and, as a result of that love, desires our happiness above all else. The Buddha (the supreme *bodhisattva*) expresses infinite love by shouldering our woes and guiding us toward *Nirvana* (ultimate peace and happiness), by visiting the Buddhist hells, and by pointing the way toward salvation. Hinduism's Krsna expresses love by freeing devotees from the cycle of karmic rebirth and so ensuring eternal happiness. The Christian God loves without end, sacrificing his son, Jesus Christ, so we might dwell, eternally happy, in paradise. Such descriptions assume a parallel between the divine and human minds. They assume that the Divine loves as we do, that the Divine feels compassion and protectiveness toward the beloved, as we do. They also assume that, like a responsible and anxious parent, the Divine possesses the power and volition to make life better for us (the children).

Spiritual happiness reflects morality, with several traditions emphasizing the spiritual value of virtue. Though virtue may take different forms (intellectual speculation, prayer and inner contemplation, generosity, hard work, detachment from worldly pursuits and desires, universal friendliness), the message remains the same: living virtuously in the here and now cashes out in terms of our future happiness. These thoughts again reflect the idea of a Divine sharing in the human moral perspective. We humans reward virtue and achievement. Academic excellence may merit prizes and research grants, extreme bravery may secure a medal from the president or the queen, hard slog and commitment may earn a promotion at work. So it is with (our projected image of) the Divine. Virtues attract divine reward; vices don't.

We believe not only that the Divine rewards us with happiness but also that the Divine (like us) experiences happiness. Happiness comprises a property of the Divine. Wishing something on behalf of other people suggests the ability to empathize with them, to feel and experience as they do. The Divine wishes for our happiness, so, we reason, the Divine must understand our situation, must share in our experience of happiness and sadness. The reality of an existence beyond our current lives, embraced by so many traditions (the Christian paradise, Islam's *Al-janna*, the Buddhist *Nirvana*), reinforces this idea of divine nature. Peace, happiness, and joy become integral to spiritual descriptions of the life to come, and because the Divine permeates all reality, these properties must pertain to the Divine.

REASONS FOR DIVINE PROJECTIONS

Why do we persistently picture a Divine who experiences and feels as we do, who shares our purposes, desires, and motivations? What function does our depiction perform? Such imaging helps insulate us from the knocks that life can deal. As children, we receive protection and nurture from our parents, older siblings, and extended family, but once grown, we must take over the role of carer for ourselves, our partners, and offspring. Sometimes we feel the need for outside support and guidance, and the image of a personal, loving, empathetic Divine can help answer this need. We feel that someone is on our side, looking out for us, easing us through the difficult times. We feel that the Divine exercises control where we lack power, that the Divine imparts meaning to apparently random, valueless events—the sudden death of a loved one, for example, or the devastating loss of life caused by famine, earthquake, or flood. Perhaps most important, we believe that our struggle and strife will ultimately receive recognition. Reward awaits us in the life to come.

Such projections may comfort us, but we should exercise caution when approaching questions about their reality. Our projected images of the Divine belong to a time well before our own. They come from a time before the rise and explanatory successes of science, to a period in history when supernatural explanations presented the best way to understand a seemingly unpredictable and often cruel world. Science has now taken over and apparently sweeps away the need for divine direction. Physics can explain the emergence of our universe in terms of the big bang; evolution and natural selection describe the variety and continued development of species, from amoeba to humankind; evolutionary psychology explains the motivation behind morality and altruism; behavioral genetics and neurochemistry describe the biological processes underlying love and happiness. We can, given the speed at which scientific research progresses, anticipate clarification of similar "mysteries" previously glossed over in supernatural or divine terms. Psychologists Steven Reiss and Susan Havercamp of Ohio State University have recently published novel research indicating that fifteen core desires (including honor, power, and human contact) drive our behavior and that our genes root almost all fifteen of them. "These desires are what guide our actions," explains Havercamp. "In a sense, we are studying the meaning of life."[3] Meaning or purpose represents yet another category that, up until now, we have treated as spiritual rather than biological territory. Modern science, therefore, undermines supernatural explanations as the only or best way to understand the chaos of a cruel world.

Do we want to lose sight of the Divine altogether? Does the success of science mean that we must abandon the Divine like a plaything we once loved but have now outgrown? No. A sense of the spiritual has, in all its myriad forms and traditions, accompanied us since time immemorial. It still accompanies us. Many of us moderns shun the traditions of organized religion, or we live the majority of our lives in an areligious way. Yet still we retain a sense of wonder and purpose. We might marvel at the beauty of a summer sunset, or perceive meaning in a worthwhile and fulfilling job, or envisage a duty to respect and protect the environment and its resources. Our goals and visions emanate beyond ourselves. They reach out to the universe as a whole. Our spirituality finds expression in an appreciation of the larger things in life. Neither does science invalidate spiritual interpretations. Accepting scientific research needn't involve drawing wholly deterministic conclusions, as the previous investigations of love and happiness have shown. Science doesn't provide a complete story; room remains for subjective, environmental, and experiential influences. The challenge becomes one of unifying both sides of the story.

RECONSTRUCTING DIVINE IMAGES

How, then, might we reconstruct our image of the Divine? How might we do justice to both scientific discovery and spiritual legacy? We need to adopt new imagery that accepts rather than challenges or ignores scientific evidence concerning the emergence and continued development of the universe, concerning the origin and driving forces of human actions and characteristics.

To take science seriously skews beyond recognition a personal vision of the Divine as usually enumerated. If our behavior nestles in our biology, we can no longer expect the spiritual, nonbiological Divine to behave as we do, to follow purposes as we do, to feel emotions as we do, to accept a morality like our own. The Divine possesses no hormones, no neurotransmitters, no genes, no veins, no cells, no blood. The Divine lacks a biology. We must rebuild our understanding of divinity from scratch.

THE DIVINE AS THE UNIVERSE-AS-A-WHOLE

I offer this novel image of the Divine: the Divine comprises the universe-as-a-whole, the totality of all that exists.[4] The Divine existed way back when that point of unimaginable heat and density began to expand—the Divine banged right along with the universe—and accompanied the universe throughout its history of development. The Divine accompanied the creation of every planet in every solar system in all the universe's galaxies and has attended the birth and gradual fading of every star. The Divine accompanied the earth through all the various ice ages and interglacial periods, through the splitting of the supercontinent Pangaea around two hundred million years ago, through the formation of the great oceans. The Divine accompanied the process of evolution as new species arose, as old ones adapted or faded away. The Divine accompanied the emergence of *Homo sapiens*, watched us develop from hunter-gatherers living hand-to-mouth into modern, urban, technological beings who dare to envision and shape our future lives. The Divine held our mother's hand as she gave birth, helped blow out every candle on all our birthday cakes, cried when our loves left us and our close ones died, and sweated with us in our illnesses. The Divine will continue to accompany the universe on its journey, wherever that may lead—perhaps toward the demise of *Homo sapiens* and the emergence of a new dominant species, perhaps ultimately toward the end of the big bang expansion, whatever form that end might take.

Out goes the old idea of a Divine distinct from the universe yet able to influence and act within it. Out goes the separation and dualism that traditional projections project. The Divine instead becomes one with the universe, diffusing throughout its parts, enveloping the whole. As the universe unfolds, so too does the Divine. As the universe develops, so too does the Divine. We identify the Divine with the substance of the universe and with its processes as the dynamic, ever-changing universe-as-a-whole.

THE DIVINE EXCEEDS THE UNIVERSE-AS-PARTS

You might hesitate to accept this offer. Surely the Divine comprises more than the universe-as-a-whole? Doesn't the Divine deserve better than being just the universe? Yes, in a sense. The universe-as-a-whole is more than the universe. Any whole exceeds the sum of its parts. Though the parts make up the whole, though the whole can't exist in the absence of its parts, still the whole surpasses its parts. The parts acting together produce a macro entity that operates to some extent independently of its components. So it is with the Divine. The parts constituting the universe together produce a macro entity, the Divine as universe-as-a-whole, that operates to some extent independently of its components.

Imagine a flock of geese migrating for winter, and then envisage how that flock behaves. All the birds fly in a tight, regimented formation, and that formation takes a life of its own. On encountering an obstacle, the flock automatically parts, merging and reforming once the coast becomes clear. Yet the flock doesn't have a lead bird directing the other geese, keeping an eye on their movements, calling them back into line if they lose position. How do the geese keep shape? How do they part and reform when the need arises? The flock (or whole) surpasses the individual geese (or parts), influencing them, taking control. Somehow the flock assumes an identity that depends on but independently exceeds the identities of the individual geese.

For "the flock" read "the Divine," and for "the individual geese" read "the universe-together-with-all-its-entities-and-processes." The Divine depends on yet somehow exceeds everything that exists. The Divine can, just like the flock, direct or exercise control over everything that exists. We must tread carefully here, however. Certainly the flock controls the movements of the individual geese, but those same individuals limit the power and capability of the flock. The flock can't make the geese fly to the moon. Neither can the flock successfully direct the geese to fly without extending their wings. The Divine also exercises limited power. The Divine can't, for example, direct the course of evolution with a future purpose in mind, contrary to the stories that Philip Hefner and Ted Peters tell. Evolution represents a random process, a response to chance events, and so its very nature precludes divine or anyone or anything else's direction. The parts feed into and influence the whole, just as the whole feeds into and influences its parts.

CONCLUSION

We hold the beginnings of a new understanding of the Divine in our hands. It respects and accepts the findings of science, the medium through which we make sense of and successfully manipulate our world. It also unifies science and spiritual insight because it not only admits that these two disciplines coexist but also shows how their subject matters (natural entities and natural processes plus the Divine) intermingle and intertwine, how they interact and influence one another. This allows room for our scientific and our spiritual commitments to take root.

This reconstructed image of the Divine won't appeal to everyone mostly because it declines the idea of a personal, superhuman figurehead. Some may feel this destroys the Divine as comforter, healer, and guide. I agree, it does demote that image, but I aim to inform, to push our horizons forward and not to continue superstitions and fears. Modern science doesn't accommodate traditional conceptions of the Divine. I don't want to abandon spirituality, however, and remain silent in the face of adversity. Rather, I want to ascertain what modern science can accommodate and rebuild a stronger, more coherent, more scientific spiritual thinking—an adequate spiritual understanding in tune with our modern age.

CHAPTER 8

Scientific Hypotheses

> [Spiritual thought] must wrestle with the best human knowledge available in the historical epoch in which a [spiritual thinker] writes. This admission . . . arises . . . out of a concern for the wrestle of [spiritual thought] with truth.[1]

I said at the end of the previous chapter that I want to build a "more scientific" spiritual understanding. I wrote this comment with serious intent. Recall the results of that study on judging risk: British people trust the word of scientists 59 percent of the time and that of religious organizations only 22 percent.[2] This finding depicts a major threat to spirituality. People view spiritual understandings as outdated, out of touch with reality, a feeble fantasy rather than a reflection of the universe's innermost nature. To rid itself of these negative connotations, spiritual thought must change; it must adapt itself to our modern scientific era. Only then may it offer its ideas as hypotheses worthy of scientific consideration.

I've already proposed a reconstructed image of the Divine, an image that coheres with our scientific understanding of the universe. How might I take my proposal further? How might I put spiritual claims to the scientific test?

OBJECTIVE SCIENCE

We rely on science every day of our lives. We watch antibiotics rid our bodies of disease, we understand that scientists can successfully make clones of farmyard animals, and we depend on up-to-the-minute technology to transmit our e-mails and faxes. We assume that, because it works, science uncovers truth. This kind of thinking originated back at the time of the scientific revolution in the seventeenth century, an era that produced great scientists like Galileo Galilei and Isaac Newton, an era that viewed empirical science as the way forward for humanity. Philosopher-of-the-time Francis Bacon wrote that that way forward involves amassing facts through careful observation and then deriving theories from those facts.

This kind of approach finds contemporary support in our commonsense conception of the aims and practices of science. Science, we assume, centers on observation, on observations available to anyone who cares to take a look. Scientists painstakingly record such observations over a period of time and under a number of different conditions. They then pass from the specific to the general, using their observations to formulate theories and universal laws. Once established, theories and laws allow scientists to make predictions and explain how things will behave in the future. Take, for example, an imaginary new treatment for rheumatism. Researchers run long-term trials: they test the new drug on patients suffering from different stages of the disease, they test for possible placebo effects, and they test the drug for harmful side effects. If at the end of the trial period the drug appears to cure various stages of the disease, its success outstrips that of placebos, and side effects seem minimal, it will appear on the market as the new wonder cure for rheumatism.

This sounds straightforward and carries intuitive appeal. We understand science as proven knowledge, and we want to rest assured in the objectivity of science, in the reliability and preference-free nature of our scientists. Such a depiction also lends support to the dualism that earlier chapters examined. On the one hand, we have the "real" sciences: objective enterprises dealing in facts, observation, and hard evidence. On the other, we have spiritual thinking and the social—the so-called soft-sciences: subjective enterprises dealing in values, personal experience, and subjective interpretation. Science we can trust, but spiritual thought and sociology we should take with a large pinch of salt.

These views received much publicity with the logical positivists (a study group formed in Vienna in the 1920s) who believe that only verification by experience can confer truth and meaning on theories and beliefs. Positivists dismiss the spiritual as meaningless because, they argue, experience can never justify spiritual belief. We can verify and so confer truth on the theory that regular exercise contributes to lengthened life span. But we can't verify our belief that, on dying, we'll rise up to Heaven and sit at the right hand of God. Richard Dawkins introduces his book *The Selfish Gene* by claiming that evolutionary science renders meaningless all other attempts to answer the questions, Why are we here? and What is our purpose? Spiritual beliefs lie beyond the pale as far as truth goes.

Many believe that the objective–subjective divide is absolute, but is it? Is it even accurate? Over the past forty years, in-depth analyses by historians and philosophers of science have cast doubt on dualist (or positivist) interpretations. A different understanding of science emerges.

SUBJECTIVE SCIENCE

THEORY AFFECTS OBSERVATION

The counterattack begins with the notion of observation. Recall that, according to our commonsense idea, scientists begin by faithfully recording publicly available observations and then go on to propose theories and universal laws on the basis of those observations. Can we rely on the objectivity and universality of our perceptions? Many philosophers and historians of science, like Thomas Kuhn and N. R. Hanson, think not; they argue for the theory-laden nature of observations. Whenever we observe, whenever we describe what we perceive, we do so from within a particular perspective or tradition. We can't observe or describe impartially, devoid of all theory. Suppose, glancing out of our window, we see a cat lying asleep on the roof of a shed in the garden. This may look like a straightforward observation, but it presupposes a lot. We need to recognize cats and garden sheds, to understand that they represent separate entities; we need to know that cats are living creatures who can sleep, wake, and change position; we need to understand the idea of an object resting on top of another object; we need to know that glass is a translucent material that bars our way but not our sight. Observation, by its very nature, requires a perspective on the world.

Not everyone shares all perspectives in common. Experimental psychologists use a piece of equipment called an eye tracker to record subjects' eye movements during experiments. Most experimental psychologists would recognize an eye tracker, knowing how and why experimenters use the machine. Yet most of us would fail to recognize the eye tracker as an eye tracker; we wouldn't know how the machine works, who uses it, or why people use it. We lack the perspective necessary for recognizing or describing this piece of equipment.

The theory ladenness of observation extends to the scientific community as well. A scientist who describes an observation does so from the perspective of a particular theory. Linguist George Lakoff illustrates this with the everyday example of a chair. From the molecular perspective, a chair equates with a collection of molecules, but from the perspective of wave equations in physics, the chair exists only as waveforms. Both statements accurately describe the chair; they just result from different scientific perspectives on or theories about the world.

THEORY DEFINES OBSERVATION

Theory plays another, even greater role within science: scientific theories guide their adherents toward observing or concentrating on particular characteristics to the exclusion of others. Modern medicine defines diseases etiologically (by their causes), explains Ludwik Fleck in his monograph *Genesis and Development of a Scientific Fact*. This etiological definition, as currently accepted, means that medical researchers study known diseases, formulate medicines and vaccinations, and isolate new diseases always with causes in mind. The etiological approach guides their mode of thinking about disease. Yet, Fleck insists, medicine could equally define diseases on the basis of other criteria such as disease symptoms. This alternative symptom-based definition, if reigning, would mean that medical researchers conduct their research always with symptoms in mind, the symptomatic idea guiding thought about disease. It might, in some instances, produce better medicine. For instance, the current approach, with its singular emphasis on causes, might miss that two causally distinct diseases share similar symptoms. Shared symptoms may mean that the same (or related) drugs could cure both diseases. In such cases, the current cause-based approach—which guides medical researchers toward observing or concentrating on particular characteristics to the exclusion of others—might impede scientific progress.

The chair and the disease examples also cast doubt on our commonsense idea that science reflects the unique truth. We assume that a theory or statement either attains truth (reflects reality) or doesn't attain truth (fails to reflect reality). Our examples indicate otherwise. Is a chair a collection of molecules or does it comprise waveforms? Should we correctly define AIDS by its cause or by its symptoms? Or, to take an example from contemporary biology, do species comprise groups of interbreeding populations (as evolutionary biologists insist), or do they comprise groups whose members share a common ancestry (as cladists insist)? Both possibilities attain truth in each example. Chairs comprise both molecules and waveforms. We can class a disease both by its cause and by its symptoms. Both groups of interbreeding populations and groups sharing a common ancestry count as species. Scientists can truthfully describe or theorize about reality in more than one way, and the way they choose reflects one possible perspective. Reality is more complex than our commonsense ideas allow, and we shouldn't expect simplicity from a science that describes that reality.[3]

KUHN'S DEPICTION OF SCIENCE

Kuhn, in his famous book *The Structure of Scientific Revolutions*, formalizes many of these thoughts, leading a revised account of the way scientists do science. Kuhn shows, on the basis of historical evidence, that science progresses not by a gradual accrual of more and more evidence either verifying or falsifying a theory about the world, but through paradigms and revolutions. Scientists within a research tradition adhere to a set of theoretical assumptions plus laws and techniques governing application of those assumptions—the set collectively called a *paradigm*—that guides their research and focuses their observations. We might consider the etiological idea of disease a scientific paradigm. If a puzzle arises that researchers can't solve within the paradigm (perhaps the identification of the cause of a specific disease doesn't help formulate a cure for it), the puzzle doesn't falsify the paradigm but becomes an *anomaly.* Scientists call the paradigm into question only when anomalies become excessive or very serious (for example, the identification of a particularly virulent and fatal disease but a cure not found despite much searching). Crisis ensues, and eventually a new paradigm emerges that covers all the previous ground, solves the troublesome anomalies, and predicts novel results. A scientific *revolution* occurs.

Kuhn can explain much of science in terms of sociology and psychology. Students become inducted into the reigning paradigm by solving standard problems, performing standard experiments, and perhaps eventually by conducting a piece of research under the supervision of an expert within the paradigm. The format of this training means that scientists remain unaware of the precise nature of the paradigm within which they work. Successful scientists don't criticize their paradigm; its theories and methods become second nature. Thus, when anomalies arise, scientists try hard to hold on to their paradigm. They may create auxiliary hypotheses to explain an anomaly, or they may claim that the anomaly represents a misinterpretation, hoping that one of their members will come up with counterevidence. When a new paradigm emerges, scientists change their allegiance for varying reasons. They may view the new paradigm as simpler, more elegant, and more cogent; they may envisage the new paradigm satisfying a pressing social need; they may feel that the paradigm solves a particularly thorny problem; or they may feel persuaded by their colleagues to accept the inevitable. Dissenters find themselves and their outdated views excluded from the scientific community once the majority of scientists have switched allegiance to the new paradigm. The revolution is complete. The old paradigm dies with the death of those who refuse to change.

A JOINT FRAMEWORK

Out goes a degree of our commonsense idea of science as an objective, rational procedure that open-minded, unbiased practitioners conduct. Enter a degree of the idea of science as subjective, irrational, blinkered, and partisan. A holistic, more rounded vision of the scientific enterprise emerges. Along with the arrival of this view dissolves the positivist distinction between verifiable science and meaningless spiritual thought. Perhaps, after all, science shares something in common with spiritual knowledge.

This Kuhnian backdrop of science as a fluid, complex, human activity encourages me to state that scientific and spiritual claims can and should be compatible, each discipline interacting with and informing the other. Room still remains in the scientific picture for experience, environment, and free will. We saw, in previous chapters, that genetic and biochemical accounts of love and happiness fail to paint a complete picture. Genes account for only 80 percent of our long-term happiness, according to Dean Hamer. Oxytocin only partially determines loving behavior, insist Thomas Insel and Sue Carter. Room still remains for spiritual perspectives on love and happiness.

AN OVERARCHING EMPIRICAL FRAMEWORK

How, then, might we present spiritual claims about love and happiness? How might we persuade scientists to accept the explanatory power of spiritual thought? Both scientists and spiritual thinkers should adopt an overarching empirical framework that makes possible the evaluation of both scientific and spiritual hypotheses concerning the nature of love and happiness. Note that experimental results begin their lives as scientific hypotheses and predictions, spoken from within theoretical traditions, based on observations that those traditions suggest. These hypotheses initially stand on the same conceptual footing as theoretical spiritual claims. I simply suggest putting the spiritual claims to the empirical test by framing them as scientific hypotheses.

The success of a joint empirical framework depends on both scientists and spiritual thinkers focusing on open, interdisciplinary approaches to ideas of common interest, admitting that spiritual and physical explanations are not mutually exclusive. Both sides must accept that any complete explanation of the nature of love or happiness will comprise spiritual as well as physical aspects since all human traits involve environment, circumstances, and individual willfulness as well as genes and biochemicals. Spiritual thinkers are thus justified in claiming that traits like happiness involve more than genes, electrical activity, and chemicals. Spiritual thinkers must, however, admit to a this-worldly nature of their claims and couch them in a form that science may evaluate.

EVALUATING SPIRITUAL CLAIMS SCIENTIFICALLY

The question becomes how or by what means we might scientifically evaluate spiritual claims about the nature of love and happiness. I can't provide a complete answer—the question is complex and this type of research in its infancy—but I can suggest several pointers toward a solution.

This book has stressed that research on the genetic and biochemical bases of human behavior is moving further and further away from unqualified reductionism. While our genetic inheritance to some extent determines the ways in which we behave, attention now focuses on other factors that combine with genes to produce a more complete explanation of our approach to and interaction with the world. Attention focuses especially on the roles of environment and culture.

Some evolutionary psychologists propose that genes and the sociocultural environment together determine our social behavior. The research mounts, and new developments now emerge in support of a joint role for genes and environment.

Social interaction influences human behavior, according to a long-term study of preschoolers by psychologist Grazyna Kochanska. Fearful children become conscientious if they receive gentle discipline based on encouragement rather than coercion, whereas fearless children ignore gentle discipline and become conscientious only when they share a cooperative, loving, and secure relationship specifically with their respective mothers. The development of conscience depends, the study thus suggests, both on the children's natural approach to the world (their genetic inheritance presumably governs this) and on specific parental practices.

Behavioral geneticist Robert Plomin strengthens support for a genetic-with-environmental approach. "We create an environment that's correlated with our genetic propensity," he says. "If that's the case, little genetic differences can push us in different directions and become larger as we go through life."[4] Someone with a natural gift for sport, for example, may seek out an environment featuring sport and other sports people. Perhaps they hang out at the gym or swimming pool. This sporty environment stimulates the relevant sporty genes and so further increases that individual's natural sporting ability. Our genes direct us toward particular environments, and in return these environments increase the behavioral effects that our genes produce.

Scholars previously construed environment and culture as nonscientific and fuzzy ideas difficult to define and evaluate: the domain of subjective humanities rather than of objective science. Now some scientists accept them as valid, evaluating and using them in scientific explanations of behavior. This could happen in the same way with potential spiritual contributions to our understanding of such behavioral traits as love and happiness; after all, as Donald Campbell and Ralph Burhoe point out, spiritual thinking forms part of our cultural heritage and social environment. Spiritual thinkers could perhaps offer hypotheses to the scientific community and scientists could evaluate those hypotheses, or scientists could rummage through their impressions of spiritual wisdom for useful hypotheses. Such scholarship can commence only in an atmosphere in which the rigid barrier between science and spiritual thought has dissolved.

RESEARCH QUESTIONS THAT MIGHT ARISE

Many questions surface regarding the relationship between spiritual and genetic accounts of human happiness. Does faith (and so, presumably, happiness) increase with involvement in a spiritual environment? Is our personal set range for happiness an indicator of how spiritual a person we are? Spiritual people declare themselves happy more often than nonspiritual people; do the set ranges of spiritual versus nonspiritual people reflect this? Do happy spiritual individuals tend to breed happy spiritual offspring? Can science make any sense of spiritual claims about happiness to come in the afterlife? For instance, might we equate our children's happiness with our own happiness in the hereafter?

Other questions arise over the relationship between spiritual and biochemical accounts of the nature of love. Do highly spiritual people exhibit greater nurture and care for their partners and offspring than do nonspiritual people? Could we establish a correlation between spirituality and high levels of oxytocin in the brain or bloodstream? Do those children with loving, spiritual parents become in turn loving spiritual parents? Can we make the decision to mend our ways and show more affection toward our partners and offspring and then act on that decision?

Questions such as these lie open to scientific evaluation and testing. They represent the road toward a more unified understanding of scientific and spiritual perspectives on love and happiness.

CONCLUSION

The implications of the holistic approach to love and happiness can flow in both directions. Spiritual thought must squarely face science, adapting and recasting itself in the light of scientific discovery. Science, likewise, must recognize its own limitations, carefully considering the insights that spiritual thought offers. The understanding of science as theory-laden and sociologically and psychologically driven allows spiritual thought a chance to offer its claims to science as hypotheses for evaluation.

Scientists' turning their attention toward the roles of environment and culture allows spiritual thinkers to enter the arena of explanation rather than loitering, forgotten, in the wings. Spiritual thinkers could make the most of these openings. Here lies the opportunity for spiritual thought to reassert itself, regain credence in the public's eyes, and, more important, help create a more healthy and holistic understanding of our world and our life.

INTERLUDE 5

Happiness, Love, and the Divine

> Neither story can say everything that there is to be said; we need both.[1]

The past couple of chapters developed a picture of a divinity at odds with traditional ideas and projections. The discussion took science seriously, encouraging spiritual thought to accept and work with science's discoveries. It also adopted, to complete the circle, an antireductionist view of science that encourages science to take spiritual thought seriously and to evaluate spiritual theses. Further, it developed a framework within which science and spiritual thought can work together to produce holistic and unified theories about our world.

LOVE, HAPPINESS, AND THE DIVINE

How we think of God, the Divine, is how *we* think of the Divine: we project what we think onto reality. Certain criteria guide such projections, a subject I won't elaborate on here, but still fundamentally the point remains the same. So how might we view God? We may choose to view God in a personal way, and in fact this may offer the best type of model or projection since we are humans and we probably want God to stand for such human ideals as justice. Does God act as a person? If we choose to interpret what we see as God's actions this way, then, yes, God acts as does a person or similarly to how a person acts. The point has to do with how we view the universe and what works best for humanity.

But is the Divine really personal? Is the universe-as-a-whole really like a person? In a straightforward way of understanding this question, probably not. But this issue does raise the one thread that remains trailing from the discussion so far. I suggested a more adequate projection of the Divine. I argued that human love and happiness comprise both biological and spiritual phenomena. But I haven't explained as yet how love and happiness relate to the Divine. The time has come to complete my handiwork.

Spiritual thinking tends to attribute to the Divine the experience and understanding of both love and happiness. Believers of many persuasions believe that the Divine feels infinite love toward us humans, desiring—above all else—our happiness. This love drives the Divine to facilitate our passage toward happiness in both this life and the next. I insisted, though, that the Divine can't experience our hopes, expectations, experiences, or sensations. We humans are biological beings, and the Divine isn't. The mistake comes from projecting intrinsic human qualities onto the nonhuman Divine. How, then, might we make sense of a nonbiological being experiencing or relating to biologically rooted, adaptive emotions like love and happiness?

LOVE AND HAPPINESS WITHIN THE WHOLE

The key lies with our new understanding of the Divine. Recall the scenario. The Divine comprises the universe-as-a-whole or the totality of everything that exists. The Divine accompanied the universe throughout the history of its development and will continue to do so. The universe and the Divine unfold together, as one. Yet the Divine exceeds everything that exists, as any whole exceeds its parts. The universe feeds and breathes life into the Divine, while the Divine drives and enfolds everything that exists.

How do love and happiness fit into this unified picture? An analogy may help illustrate the answer. Take yourself back to when John Kennedy was president of the United States. The Democratic Party then possessed a spirit, a system of beliefs, and a life that enfolded but transcended the spirit of Kennedy, his system of beliefs, and his life, as well as those of all its other members, past, present, and future. The spirit of Kennedy helped form the spirit of the Democratic Party. In particular, Kennedy probably experienced happiness when first elected his party's presidential candidate. As a husband and father, he no doubt felt love toward his wife and children. These happy and loving facets of Kennedy became, along with the man, part of the Democratic Party. We don't say that the Democratic Party experienced love and happiness as did Kennedy. A political party isn't the kind of thing that can experience these emotions; it lacks the biological background—a physical body filled with genes, hormones, and biochemicals—necessary for such experience. Kennedy's experience of love and happiness, however, probably influenced his party. Perhaps the love he felt toward his wife and children helped mold his opinions about divorce, single parenthood, or education. Perhaps they informed his attitudes toward civil rights. Democratic Party policy reflected these opinions. Perhaps Kennedy's delight at election as president caused him to lead his party wisely and well in the hope of personal reelection. The party thus absorbed Kennedy's love and happiness, these emotions shaping and sustaining the party, adding to its history, enlarging its spirit.

In a similar way, as we form parts of the universe-as-a-whole, of the Divine, so our love and happiness also help form the loving and happiness properties of the Divine. Our feelings of love and happiness shape and sustain, add to and enlarge divine love and happiness. We must, of course, view this divine love and happiness differently from ours because they lie in the context of the whole. The Divine lacks the biological background necessary to experience love and happiness as we do.

Influence flows in the opposite direction too because the whole—the party, the Divine—pervades and colors its parts. Kennedy's views on divorce, education, single parenting, and civil rights may have sprung not only from love for his family but also from Democratic ideals concerning human rights, social justice, and individual responsibility—perhaps the ideals that initially drew him to the party. The wisdom in Kennedy's leadership probably reflected his belief that those ideals represent the best way forward for the people of the United States. The party thus created something more of Kennedy's political beliefs, something that reflected the history of the party, its stated ideals, and the shared attitudes of everyone in it. It transcended Kennedy.

Similarly, divine love and happiness include but go beyond our experience of love and happiness. They relate intimately to our experience but exceed it in a holistic way as any whole both embraces and transcends its parts.

We talk about the spirit of the Democratic Party. By this we mean a nonstatic, fluid conglomerate of ideals, attitudes, goals, and beliefs that encapsulate yet transcend the beliefs and actions of individual Democrats. So we can speak of divine love and happiness as nonstatic, fluid conglomerates that encapsulate yet transcend the love and happiness of individual human beings.

QUESTIONS AND OBJECTIONS

Important questions remain unanswered. Under the analogy I've provided, how can the Divine love us or desire our happiness? A political party isn't the kind of thing that can experience love or longing.

Perhaps the answer again lies in the way the whole comprises more than its parts, in the way the Divine enfolds our human desires. The Democratic Party possesses a past, a present, and a future. Past events and individuals helped shape the spirit of the party. Present events lead on from past events, and present members look backward to the party's history to seek out values, examples, and precedents. Present events and individuals will shape and influence the future of the party. So the cycle continues.

Perhaps we, too, look backward to past events and past individuals in search of guidance, values, and examples. Past experience teaches that tobacco, hard drugs, and alcohol may lead to superficial highs but will ultimately result in great unhappiness and disrupted lives. Couples who fight may attend marriage guidance sessions where they can draw on the experience of others to fortify and rebuild their flagging partnerships. Our own experiences, in turn, enter the holistic stockpile and may provide fodder for future generations. Happiness and love draw us; we desire them for ourselves. The Divine enfolds our desires and, by furnishing us with a spirit, with a vista of past experiences and situations, enables us to make judgments and construct values for our future. We learn how, with the help of this vista, to satisfy our desires successfully; we learn how to attain happiness and form loving relationships for ourselves.

What of the afterlife—that eternity of ultimate happiness—which, so many spiritual traditions suggest, awaits us? Does this have a place in our universe-as-a-whole? If it does, its form may differ from some traditional ideas. Scientific descriptions leave little room for a paradise filled with disembodied spirits of the great and good. We yearn for comfort and reward, and we specialize in projections. Does a bounteous afterlife represent yet another projection? Perhaps, if we do want to affirm the existence of an afterlife, we should take a leaf from the Confucian book and treat an individual's life as continuing those of his or her ancestors, a later link in the family chain. Could the afterlife reside in our children's unfolding lives? We certainly pass our genes on to our offspring; they even inherit many of our behavioral traits. As a second possible vision of the afterlife, we might think of ourselves as living on through the legacy of our experiences and actions, with others building on or learning from our successes and failures. These pictures of the afterlife, though, lack the bounty and delight of the Christian paradise or the Islamic *Al-janna*.

Many scholars disagree with this reconstruction of the Divine, with this understanding of love and happiness. They would prefer a transcendent Divine who exists distinct from the universe. They prefer to view love and happiness as divine properties that we humans have gradually evolved into; the Divine had them, and we evolved them by chance or by divine design. To these points, my stands remain the same:

- Evolution proceeds with naturally produced variations subject to local conditions and changes; this happenstance nature of evolution provides no evidence of divine purpose or foresight. Science doesn't support pointing fingers and booming celestial voices, the stuff of divine intervention.
- Love and happiness are characteristics that we have evolved in response to biological needs and environmental pressures. The protagonists surely don't think their God crossed some fingers and hoped for us to develop the right way? How can we evolve the same sort of properties as a nonbiological being so different from ourselves and in a quite different environment? Human love and happiness as products of evolution would surely vastly differ from the divine equivalents, if they exist. We don't expect the Divine to possess a digestive tract just because we evolved to have one.

- Furthermore, a universe-distinct Divine invokes the kind of scientific–spiritual dualism that I argue against because, for one thing, it fails to tally with the biology of our subjective selves (consider the biochemistry and genetics of love and happiness). Little also remains for a universe-distinct Divine to do; for instance, intervention is out, and science denies that the big bang equates with creation out of nothing (some kind of primitive logic or geometry preexisted). The protagonists' position leads to inconsistency and confusion.

CONCLUSION

Love and happiness lie entwined at the heart of our lives. Well-being represents the most common goal toward which we strive, and loving relationships frequently aid us in finding that happiness. I sought to understand and explain their significance because of the centrality of these emotions to our everyday lives. My picture emphasizes their biological rootedness while opening the door to free will, subjectivity, and spirituality.

Our journey of discovery led beyond love and happiness to the wider dispute between evolutionary psychology and spiritual thought, between science and spiritual thought. I engaged in serious reconstruction: a renewed understanding of the Divine, a recasting of science and spiritual thought that allows scientific evaluation of spiritual theses, an alternative conception of love and happiness that complements our renewed understanding of the Divine.

Traditionalists may feel dismay: I've overhauled so much. I insist that the overhaul remains necessary. Adherence to tradition will ultimately lead to its death in a world of science and a world even more bereft of value and wisdom than at present. Instead, I strive for a united system of scientific and spiritual thinking, for a system that leads to deep, holistic, satisfying explanations, a system that does justice to both the biological and the spiritual sides of our nature. As the opening quotation explains, "Neither story can say everything that there is to be said; we need both."

Notes

CHAPTER 1

1. Kaplan 1951: 80–83, 301.
2. van Biema 1997: 72.
3. van Biema 1997: 77.
4. Myers and Diener 1996.
5. Wu 1992: 31, quoted in Luo and Shih 1997: 183.
6. Qur'an, Sura 91. 7–10, quoted in Hinnells 1997: 176.
7. Qur'an, Sura 21. 35–36, quoted in Hinnells 1997: 176.
8. Kohler 1968; Mattuck 1953.

CHAPTER 2

1. Murdoch 1993: 17.
2. Myers and Diener 1996.
3. Hamer 1996.
4. Howe Colt 1998.
5. Lykken and Tellegen 1996: 188.
6. Hamer 1996: 125.
7. Costa et al. 1987.

8. Lewis and Joseph 1997.
9. Grantham 1996: D2.
10. Gose 1996.
11. Diener and Diener 1996.
12. Myers 1992a: 48.
13. Grantham 1996.
14. Gose 1996: A9.
15. Goleman 1996: B9.
16. Holden 1996: 1593–94.
17. Hamer 1996: 126.
18. Goleman 1996: B9.
19. Lane et al. 1997: 930.
20. Myers 1992a.
21. Myers 1992b.
22. Myers 1992b.
23. Corelli 1996.
24. Myers and Diener 1995: 13.
25. Ruse 1992: 163.
26. Brickman and Campbell 1971: 289.
27. Brickman and Campbell 1971: 287.
28. Wright 1994: 298.
29. Pinker 1997: 390.
30. Pinker 1997: 391.

INTERLUDE 1

1. Freeman 1997: 70.
2. Marris and Langford 1996.
3. Ebstein et al. 1996: 78.
4. Bower 1996b: 279.
5. Bower 1996a.
6. Plomin 1990: 248.
7. Lesch et al. 1996: 1527.
8. Howe Colt 1998: 42.
9. Begley 1996: 79.
10. Lykken et al. 1992: 1565.
11. Begley 1997.
12. Lykken and Tellegen 1996: 186.

13. Myers and Diener 1995: 12.
14. Wright 1997: 95.
15. Wright 1997: 90.
16. Freeman 1997: 70.
17. Freeman 1997: 68.
18. Epstein 1995: 42.
19. Goleman 1996: B9.
20. Hamer 1996: 126.
21. Plomin 1990: 187.

CHAPTER 3

1. 1 Corinthians 1:1–13. This and subsequent Bible references come from *The New English Bible.*
2. Plato 1997: 202e–203a.
3. Plato 1997: 207a.
4. Augustine, *City of God* 11.28, quoted in Long 1987: 38.
5. Aquinas, *Commentary on the Divine Names,* quoted in Long 1987: 38.
6. Lehrer 1994.
7. Mark 12:29–32.
8. John 15:12–13.
9. 1 John 4:7–8.
10. Matthew 5:44–46.
11. Song of Songs 1:1–4.
12. Mother Teresa 1995: 83.
13. Confucius, *Analects* 12.2, quoted in Long 1987: 32.
14. Confucius, *Analects* 4.3, quoted in Long 1987: 32.
15. Mo-tzu, quoted in Long 1987: 32; reworded as gender inclusive.
16. *Tao-te ching,* chap. 37, quoted in Long 1987: 33.
17. Long 1987: 35.
18. *Bhagavadgita* 11.55, quoted in Long 1987: 34.
19. Glassé 1989: 357.
20. Glassé 1989: 357.
21. Glassé 1989: 357.
22. Glassé 1989: 358.
23. Kohler 1968: 31–32, emphasis removed.

24. Kohler 1968: 484–85.
25. Mattuck 1953: 54.

CHAPTER 4

1. Crenshaw 1996: 194.
2. Wright 1994: 351.
3. Hamer and Copeland 1998: 199.
4. Radetsky 1994: 1487.
5. Insel 1997: 728.
6. Ezzell 1992a: 7.
7. Carter and Getz 1993: 104.
8. Uvnäs-Moberg 1997: 154.
9. Travis 1996: 247.
10. Travis 1996: 247.
11. Travis 1996: 247.
12. Young et al. 1997b: 225.
13. Travis 1996: 247.
14. Young et al. 1997b: 227.
15. Young et al. 1997b: 227.
16. Insel 1997: 728.
17. Carter and Getz 1993: 105.
18. Insel 1997: 729.
19. Uvnäs-Moberg 1997: 155.
20. Insel and Carter 1995: 14.
21. Insel and Carter 1995: 14.

INTERLUDE 2

1. Freeman 1997: 70.
2. Crenshaw 1996: 194.
3. 1 John 4:7–9.
4. Quoted in Woodward 1998: 50.
5. Zabludoff 1997: 10.
6. Wright 1992: 38.
7. Zabludoff 1997: 10.
8. Wright 1992: 38.
9. Insel 1997: 726.

10. Freeman 1997: 70.
11. Insel 1997: 731.
12. Insel 1997: 733.
13. Wang and Insel 1996: 380.

CHAPTER 5

1. W. Somerset Maugham, quoted in McKernan 1994: 22.
2. Ruse 1989: 260.
3. Ruse and Wilson 1985: 51.
4. Ruse and Wilson 1985: 265.
5. Alexander 1987; Irons 1991.
6. Ruse 1989: 262.
7. Ruse 1989: 267.
8. Ruse 1989: 268.
9. Campbell 1976: 170.
10. Campbell 1976: 172.
11. Campbell 1976: 192.
12. Dawkins 1989: 192.
13. Burhoe 1981: 20.
14. Burhoe 1981: 20.
15. Ruse and Wilson 1985: 52.

CHAPTER 6

1. Gilkey 1989: 19.
2. Manenschijn 1988: 86.
3. Peacocke 1986: 111.
4. Peacocke 1986: 110.
5. From a coauthor interview with Hamer.
6. Wilson 1978: 167.
7. Peacocke 1986: 113.
8. Job 13:15.
9. Peacocke 1986: 114.
10. Peacocke 1986: 114.
11. Peacocke 1986: 114. Peacocke has written more recently about sociobiology in his book *Theology for a Scientific Age*. Though he employs less polemic language here, still his criticisms remain

the same: asserting that sociobiology can't wholly explain culture, upholding the distinction between origins and content, and accusing sociobiologists of the "genetic fallacy."

12. Bowker 1980: 327.
13. Bowker 1980: 328.
14. Bowker 1980: 316.
15. Bowker 1995: 5–6.
16. Bowker 1995: 8.
17. Bowker 1995: 106.
18. From a coauthor interview with Hamer.
19. Motley 1998: 10.

INTERLUDE 3

1. Hefner 1984b: 190.
2. Hefner 1989a: 212.
3. Hefner 1984b: 203.
4. Hefner 1984b: 202.
5. Hefner 1984b: 204.
6. Hefner 1980: 208.
7. Hefner 1980: 210.
8. Hefner 1989b: 314.
9. Hefner 1989a: 222.
10. Hefner 1984a: 198.
11. Hefner 1989b: 319.
12. Hefner 1989a: 211.
13. Hefner 1984a: 201.
14. Hefner 1984a: 211–12.
15. Hefner 1989a: 226, 232.
16. Hefner 1984a: 214.
17. Hefner 1989a: 212.
18. Hefner 1984a: 215.

INTERLUDE 4

1. Peters 1989a: 110.
2. Peters 1989c: 13–14.
3. Peters 1989b: xii–xiii.

4. Peters 1989b: 125.
5. Peters 1989b: 128.
6. Peters 1989b: 125.
7. Peters 1989b: 126.
8. Peters 1989b: 131.
9. Peters 1989b: 132.
10. Peters 1989b: 133.
11. Peters 1989b: 133
12. Peters 1989b: 134.
13. Peters 1989b: 135.
14. Peters 1989b: 136.
15. Peters 1989b: 138.
16. Peters 1989b: 138.
17. Peters 1989b: 136.
18. Peters 1989b: 136.
19. Sharpe 2000.
20. Peters 1994: 302.
21. Peters 1994: 303.

CHAPTER 7

1. From the documents of the Fourth Lateran General Council of 1215, quoted in Peters 1989b: 88.
2. From the Westminster Shorter Catechism, quoted in Peters 1989b: 88.
3. Patton 1998: 8.
4. See Sharpe 2000.

CHAPTER 8

1. Godbey 1970: 208.
2. Marris and Langford 1996.
3. See Bryant 1997 for an extensive discussion of this topic.
4. Holmes 1997: 16.

INTERLUDE 5

1. Watts 1997: 135.

Bibliography

Ackrill, J. L., ed. 1987. *A New Aristotle Reader*. Oxford: Clarendon Press.

Adler, J. 1996. The Happiness Meter. *Newsweek* 128, no. 5 (July 19): 78.

Alexander, R. D. 1987. *The Biology of Moral Systems*. New York: Aldine De Gruyter.

Begley, S. 1996. Born Happy. *Newsweek* 128, no. 16 (October 14): 78–80.

———. 1997. Wombs with a View. *Newsweek* 130, no. 6 (August 11): 61.

———. 1998. Science Finds God. *Newsweek* 132, no. 4 (July 27): 44–49.

Bower, B. 1994. Hormone Shows Link to Some Obsessions. *Science News* 146, no. 18 (October 29): 277.

———. 1996a. Gene Connected to Human Cognitive Trait. *Science News* 150, no. 3 (July 20): 39.

———. 1996b. New Data Challenge Personality Gene. *Science News* 150, no. 18 (November 2): 279.

———. 1997. Conscience Grows on Temperamental Ground. *Science News* 151, no. 13 (March 29): 189.

Bowker, J. W. 1980. The Aeolian Harp: Sociobiology and Human Judgment. *Zygon* 15, no. 3: 307–33.

——. 1995. *Is God a Virus?* London: SPCK.

Brandon, G. F. 1970. *A Dictionary of Comparative Religion*. London: Weidenfeld & Nicholson.

Brickman, P., and D. T. Campbell. 1971. Hedonic Relativism and Planning the Good Society. In *Adaptation Level Theory*, ed. M. H. Appley, 287–302. New York: Academic Press.

Bryant, R. 1997. Discovery and Decision: Exploring the Metaphysics and Epistemology of Scientific Classification. Ph.D. diss., University of Edinburgh.

Burhoe, R. W. 1981. *Toward a Scientific Theology*. Belfast: Christian Journals.

Campbell, D. T. 1976. On the Conflicts between Biological and Social Evolution and between Psychology and Moral Tradition. *Zygon* 11, no. 3: 167–208.

Carmichael, M. S., V. L. Warburton, J. Dixen, and J. M. Davidson. 1994. Relationships among Cardiovascular, Muscular, and Oxytocin Responses during Human Sexual Activity. *Archives of Sexual Behavior* 23: 59–79.

Carter, S. C., and L. L. Getz. 1993. Monogamy and the Prairie Vole. *Scientific American* 268 (June): 100–106.

Compton, N. 1998. Secrets and Lies (The 1998 Arena Sex Survey). *Arena* 81 (September): 73–81.

Corelli, R. 1996. Get Happy: Experts Debate whether the Key to Happiness Lies in the Genes. *Maclean's* 109, no. 38 (September 16): 54–58.

Costa, P. T., Jr., R. R. McCrae, and A. B. Zonderman. 1987. Environmental and Dispositional Influences on Well-Being: Longitudinal Follow-Up of an American National Sample. *British Journal of Psychology* 78, no. 3 (August): 299–306.

Crenshaw, T. 1996. The Alchemy of Love and Lust. *Cosmopolitan*, March, 192–95.

Csikszentmihalyi, M. 1990. *Flow: The Psychology of Optimal Experience*. New York: Harper & Row.

Da Costa, A. P. C., R. G. Guevara-Guzman, S. Ohkura, J. A. Goode, and K. M. Kendrick. 1996. The Role of Oxytocin Release in the

Paraventricular Nucleus in the Control of Maternal Behavior in the Sheep. *Journal of Neuroendocrinology* 8, no. 3: 163–77.

Damasio, A. R. 1994. *Descartes' Error: Emotion, Reason, and the Human Brain*. New York: Avon.

Dawkins, R. 1989 [1976]. *The Selfish Gene*. Oxford: Oxford University Press.

Depue, R. A., M. Luciana, P. Arbisi, P. Collins, and A. Leon. 1994. Dopamine and the Structure of Personality: Relation of Agonist-Induced Dopamine Activity to Positive Emotionality. *Journal of Personality and Social Psychology* 63, no. 7 (September): 485–98.

Diener, E., and C. Diener. 1996. Most People are Happy. *Psychological Science* 7, no. 3: 181–85.

Ebstein, R. P., O. Novick, R. Umansky, B. Priel, Y. Osher, D. Blaine, E. R. Bennett, L. Nemanov, M. Katz, and R. H. Belmaker. 1996. Dopamine D4 Receptor (D4DR) Exon III Polymorphism Associated with Human Personality Trait of Novelty Seeking. *Nature Genetics* 12, no. 1 (January): 78–80.

Epstein, M. 1995. Opening Up to Happiness. *Psychology Today* 28, no. 4 (July–August): 42–47.

Ezzell, C. 1992a. Brain Receptors Shape Voles' Family Values. *Science News* 142 (July 4): 6–7.

———. 1992b. Explanation for Premature and Delayed Labor. *Science News* 141 (June 13): 389.

Fleck, L. 1979. *Genesis and Development of a Scientific Fact*. Edited by T. J. Trenn and R. K. Merton. Chicago: University of Chicago Press.

Freeman, W. J. 1997. Happiness Doesn't Come in Bottles: Neuroscientists Learn That Joy Comes through Dancing, Not Drugs. *Journal of Consciousness Studies* 4, no. 1: 67–70.

Gilkey, L. 1989. Insights from a Senior Associate. *Newsletter of the Chicago Center for Religion and Science* 1, no. 2 (December): 19.

Glassé, C. 1989. *The Encyclopaedia of Islam*. London: Stacey International.

Godbey, J. C. 1970. Further Remarks on the Need for a Scientific Theology. *Zygon* 5, no. 3 (September): 194–215.

Goleman, D. 1996. Forget Money: Nothing Can Buy Happiness, Some Researchers Say. *New York Times*, July 16, B5, B9.

Gose, B. 1996. Seeking Genetic Roots of Contentment. *Chronicle of Higher Education* 43, no. 10 (November 1): A9.

Grantham, R. 1996. Lotta Luck? *Ann Arbor News*, December 15, D1–D2.

Greenblatt, R. B., and V. B. Mahesh. 1996. Important Human Hormones. *Collier's Encyclopedia*. Cognito article, no. 17147687.

Hamer, D. H. 1996. The Heritability of Happiness. *Nature Genetics* 14, no. 6 (October): 125–26.

Hamer, D. H., and P. Copeland. 1998. *Living with Our Genes*. New York: Doubleday.

Hanson, N. R. 1958. *Patterns of Discovery*. Cambridge: Cambridge University Press.

Hefner, P. 1980. Survival as a Human Value. *Zygon* 15, no. 2: 203–12.

———. 1984a. Creation: Viewed by Science, Affirmed by Faith. In *Cry of the Environment: Rebuilding the Christian Creation Tradition*, ed. P. N. Joranson, 198–217. Santa Fe, N.M.: Bear & Co.

———. 1984b. Sociobiology, Ethics, and Theology. *Zygon* 19, no. 2: 185–207.

———. 1989a. The Evolution of the Created Co-Creator. In *Cosmos as Creation: Theology and Science in Consonance*, ed. T. Peters, 211–33. Nashville: Abingdon Press.

———. 1989b. An Exercise in Theological Anthropology: The Created Co-Creator and Ethical Norms. In *Kooperation und Wettbewerb: Zu Ethik und Biologie menschlichen Sozialverhaltens*, Loccumer Protokolle, vol. 75/1988, ed. H. May, M. Striegnitz, and P. Hefner, 313–32. Rehburg-Loccum, Germany: Evangelische Akademie Loccum.

Hinnells, J. R., ed. 1997. *A New Handbook of Living Religions*. Oxford: Basil Blackwell.

Holden, C. 1996. Happiness and DNA. *Science* 272, no. 5268 (June 14): 1591–93.

Holmes, B. 1997. Twins Spring Gene Shocker. *New Scientist* 154, no. 2068 (June 14): 16.

Howe Colt, G. 1998. Were You Born That Way? *Life* 21, no. 4 (April): 38–49.

Hrdy, S. B., and S. C. Carter. 1995. Hormonal Cocktails for Two. *Natural History* 104: 34.

Insel, T. R. 1997. A Neurobiological Basis of Social Attachment. *American Journal of Psychiatry* 154, no. 6: 726–35.

Insel, T. R., and S. C. Carter. 1995. The Monogamous Brain. *Natural History* 104: 13–14.

Irons, W. 1991. How Did Morality Evolve? *Zygon* 26, no. 1: 49–89.

Kaplan, J. D. 1951. *Dialogues of Plato*. New York: Pocket Books.

Kochanska, G. 1997. Multiple Pathways to Conscience for Children with Different Temperaments: From Toddlerhood to Age 5. *Developmental Psychology* 33, no. 2: 228–40.

Kohler, K. 1968 [1918]. *Jewish Theology: Systematically and Historically Considered*. New York: KTAV Publishing House.

Kuhn, T. S. 1970 [1962]. *The Structure of Scientific Revolutions*. Chicago: University of Chicago Press.

Lakoff, G. 1987. *Women, Fire, and Dangerous Things: What Categories Reveal about the Mind*. Chicago: University of Chicago Press.

Lane, R. D., E. M. Reiman, G. L. Ahern, G. E. Schwartz, and R. J. Davidson. 1997. Neuroanatomical Correlates of Happiness, Sadness, and Disgust. *American Journal of Psychiatry* 154, no. 7: 926–33.

Lefebvre, D. L., A. Giaid, H. Bennett, R. Larivière, and H. H. Zingg. 1992. Oxytocin Gene Expression in Rat Uterus. *Science* 256 (June 12): 1553–55.

Lehrer, T. 1994 [1959]. She's My Girl. Song on the CD *Tom Lehrer in Concert*. London: Decca Record Co.

Lesch, K-P., D. Bengel, A. Heils, S. Sabol, B. Greenberg, S. Petri, J. Benjamin, C. Muller, D. Hamer, and D. Murphy. 1996. Association of Anxiety-Related Traits with a Polymorphism in the Serotonin Transporter Gene Regulatory Region. *Science* 274 (November 29): 1527–30.

Lewis, C. A., and S. Joseph. 1997. The Depression-Happiness Scale: A Measure of a State or a Trait? *Psychological Reports* 81, no. 3: 1313–14.

Long, J. B. 1987. Love. In *Encyclopaedia of Religion*, 31–40. New York: Macmillan.

Ludwig, M. 1995. Functional Role of Intrahypothalamic Release of Oxytocin and Vasopressin: Consequences and Controversies. *American Journal of Physiology* 268: E537–45.

Lumsden, C. J. 1988. Psychological Development: Epigenetic Rules and Gene-Culture Coevolution. In *Sociobiological Perspectives in Human Development*, ed. K. B. Macdonald, 234–67. New York: Springer-Verlag.

Lumsden, C. J., and E. O. Wilson. 1983. *Promethean Fire: Reflections on the Origin of Mind*. Cambridge, Mass.: Harvard University Press.

Luo, L., and J. B. Shih. 1997. Sources of Happiness: A Qualitative Approach. *Journal of Social Psychology* 137, no. 2: 181–87.

Lykken, D. T., M. McGue, A. Tellegen, and T. J. Bouchard Jr. 1992. Genetic Traits that May Not Run in Families. *American Psychologist* 47, no. 12 (December): 1565–77.

Lykken, D. T., and A. Tellegen. 1996. Happiness Is a Stochastic Phenomenon. *Psychological Science* 7, no. 3 (May): 186–89.

Manenschijn, G. 1988. Evolution and Ethics. In *Evolution and Creation*, ed. S. Anderson and A. R. Peacocke, 85–103. Aarhus, Denmark: Aarhus University Press.

Marris, C., and I. Langford. 1996. No Cause for Alarm. *New Scientist* 151, no. 2049 (September 28): 36–39.

Mattuck, I. 1953. *Jewish Ethics*. London: Hutchinson's University Library.

McEwen, B. S. 1997. Meeting Report: Is There a Neurobiology of Love? *Molecular Psychiatry* 2: 15–16.

McGrath, A. E., ed. 1993. *The Blackwell Encyclopaedia of Modern Christian Thought*. Oxford: Basil Blackwell.

McKernan, R. 1994. What's Love Got to Do with It? Lots. *Independent*, December 20, 22.

Mother Teresa. 1995. *A Simple Path*. Compiled by L. Vardey. London: Rider.

Motley, I. 1998. Nature vs. Nurture. *Life* 21, no. 4 (April): 10.

Murdoch, I. 1993. *The Green Knight*. London: Chatto & Windus.

Myers, D. G. 1992a. *The Pursuit of Happiness*. New York: William Morrow.

——. 1992b. The Secrets of Happiness. *Psychology Today* 25, no. 4 (July–August): 38–46.

——. 1993. Pursuing Happiness. *Psychology Today* 26, no. 4 (July–August): 32–36.

Myers, D. G., and E. Diener. 1995. Who Is Happy? *Psychological Science* 6, no. 1 (January): 10–19.

——. 1996. The Pursuit of Happiness. *Scientific American* 274, no. 5 (May): 54–56.

Nelson, E., and J. Panksepp. 1996. Oxytocin Mediates Acquisition of Maternally Associated Odor Preferences in Preweanling Rat Pups. *Behavioral Neuroscience* 110, no. 3: 583–92.

The New English Bible. 1970. Oxford: Oxford University Press, and Cambridge: Cambridge University Press.

Nissen, E., G. Lilja, A. M. Widstrom, and K. Uvnäs-Moberg. 1995. Elevation of Oxytocin Levels Early Post Partum in Women. *Acta Obstetricia et Gynecologica Scandinavica* 74: 530–33.

Nygren, A. 1953. *Agape and Eros*. London: SPCK.

Oxytocin Receptors Linked to Social Behavior. 1992. *BioScience* 42: 327.

Patton, L. 1998. Sex Low on Public List of Priorities When Seeking the Meaning of Life. *The Guardian*, June 16, 8.

Peacocke, A. R. 1986. *God and the New Biology*. London: J. M. Dent & Sons.

——. 1993 [1990]. *Theology for a Scientific Age*. London: SCM Press.

Peters, T. 1989a. Cosmos as Creation. In *Cosmos as Creation: Theology and Science in Consonance*, ed. T. Peters, 41–114. Nashville: Abingdon Press.

——. 1989b. *God: The World's Future*. Minneapolis: Fortress Press.

——. 1989c. Preface. In *Cosmos as Creation: Theology and Science in Consonance*, ed. T. Peters, 11–27. Nashville: Abingdon Press.

——. 1994. *Sin: Radical Evil in Soul and Society*. Grand Rapids, Mich.: William B. Eerdmans Publishing.

The Philosophical Works of Descartes. 1969 [1911]. Edited and translated by E. Haldane and G. R. T. Ross. Cambridge: Cambridge University Press.

Pinker, S. 1997. *How the Mind Works*. London: Allen Lane.

Plato. 1993a. *Phaedo*. Translated by D. Gallop. Oxford: Oxford University Press.

——. 1993b. *The Symposium and the Phaedrus: Plato's Erotic Dialogues*. Translated by W. S. Cobb. Albany: State University of New York Press.

——. 1987. *Republic*. 2nd ed. Translated by D. Lee. London: Penguin.

——. 1997. *Symposium and the Death of Socrates*. Translated by T. Griffith. Ware, U.K.: Wordsworth.

Plomin, R. 1990. The Role of Inheritance in Behavior. *Science* 248, no. 4952 (April 13): 183–87.

Radetsky, P. 1994. Stopping Premature Births before It's Too Late. *Science* 266 (December 2): 1486–88.

Reich, J., E. Diener, D. G. Myers, and A. C. Michalos. 1994. The Road to Happiness. *Psychology Today* 27: 32–37.

Reiss, S., and S. M. Havercamp. 1998. Toward a Comprehensive Assessment of Fundamental Motivation: Factor Structure of the Reiss Profiles. *Psychological Assessment* 10, no. 2: 97–106.

The Rider Encyclopaedia of Eastern Philosophy and Religion. 1989. London: Rider.

Rosenblatt, J. S. 1994. Psychobiology of Maternal Behavior: Contribution to the Clinical Understanding of Maternal Behavior among Humans. *Acta Paediatrica*, supplement 397, 3–8.

Ruse, M. 1989. *The Darwinian Paradigm: Essays on its History, Philosophy, and Religious Implications*. London: Routledge.

——. 1992. *Evolutionary Naturalism*. London: Routledge.

Ruse, M., and E. O. Wilson. 1985. The Evolution of Ethics. *New Scientist* 108, no. 1478: 50–52.

Sarlin, C. N. 1981. The Role of Breast-Feeding in Psycho-Sexual Development and the Achievement of the Genital Phase. *Journal of the American Psychoanalytic Association* 29, no. 3: 631–41.

Schrof, J. M. 1991. Hormone of Love: The Chemistry of Romance and Nurturance. *U.S. News & World Report* 110 (June 24): 62.

Schuller, R. 1985. *The Be Happy Attitudes: Eight Positive Attitudes That Can Transform Your Life*. Waco, Tex.: Word Books.

Sharpe, K. 2000. *Sleuthing the Divine: The Nexus of Science and Spirit*. Minneapolis: Fortress Press.

Smith, J. Z., ed. 1995. *The HarperCollins Dictionary of Religion*. San Francisco: HarperCollins.

Sobel, D. 1995. Interview: Mihaly Csikszentmihalyi. *Omni* 17, no. 4 (January): 73–80.

Steen, R. G. 1996. *DNA and Destiny: Nature and Nurture in Human Behavior*. New York: Plenum Press.

Travis, J. 1996. A Hormone's Reputation Takes a Beating. *Science News* 150, no. 16: 246–47.

Uvnäs-Moberg, K. 1996. Neuroendocrinology of the Mother-Child Interaction. *Trends in Endocrinology and Metabolism* 7, no. 4: 126–31.

———. 1997. Physiological and Endocrine Effects of Social Contact. *Annals of the New York Academy of Sciences* 807: 146–63.

———. 1998. Antistress Pattern Induced by Oxytocin. *News in Physiological Sciences* 13: 22–26.

van Biema, D. 1997. Does Heaven Exist? *Time* 149, no. 12 (March 24): 70–78.

van Erp, A. M., M. R. Kruk, J. G. Veening, T. A. Roeling, and W. Meelis. 1995. Neuronal Substrate of Electrically Induced Grooming in the PVH of the Rat: Involvement of Oxytocinergic Systems? *Physiology and Behavior* 57: 882–85.

Wang, Z., and T. R. Insel. 1996. Parental Behavior in Voles. *Advances in the Study of Behavior* 25: 361–84.

Watts, F. 1997. Are Science and Religion in Conflict? *Zygon* 32, no. 1: 125–38.

Wilson, E. O. 1978. *On Human Nature*. Cambridge, Mass.: Harvard University Press.

Winslow, J. T., L. Shapiro, S. C. Carter, and T. R. Insel. 1993. Oxytocin and Complex Social Behavior: Species Comparison. *Psychopharmacology Bulletin* 29, no. 3: 409–14.

Woodward, K. J. 1998. How the Heavens Go. *Newsweek* 132, no. 4 (July 27): 50.

Wright, L. 1997. *Twins: Genes, Environment, and the Mystery of Human Identity.* London: Weidenfeld & Nicholson.

Wright, R. 1992. Science, God, and Man. *Time* 140, no. 26: 38.

———. 1994. *The Moral Animal.* New York: Vintage.

Wurtzel, E. 1995. *Prozac Nation: Young and Depressed in America.* London: Quartet.

Young, L. J., Z. Wang, and T. R. Insel. 1998. Neuroendocrine Bases of Monogamy. *Trends in Neuroscience* 21: 71–75.

Young, L. J., J. T. Winslow, R. Nilsen, and T. R. Insel. 1997a. Species Differences in V_1a Receptor Gene Expression in Monogamous and Nonmonogamous Voles: Behavioral Consequences. *Behavioral Neuroscience* 111, no. 3: 599–605.

Young, L. J., J. T. Winslow, Z. Wang, B. Gingrich, Q. Guo, M. M. Matzuk, and T. R. Insel. 1997b. Gene Targeting Approaches to Neuroendocrinology: Oxytocin, Maternal Behavior, and Affiliation. *Hormones and Behavior* 31, no. 3: 221–31.

Zabludoff, M. 1997. Behaving Ourselves. *Discover* 18 (October 1): 10.

INDEX

INDEX

ABOUT THE AUTHORS

For most of his life, **Kevin Sharpe** has studied the nexus of science and spirituality. His academic background includes doctorates in mathematics and religious studies, and research in prehistoric archaeology. He is a Core Professor in the Graduate College of Union Institute and University, Cincinnati, and a member of Harris Manchester College, Oxford University. Of his several books, the most recent is *Sleuthing the Divine: The Nexus of Science and Spirit*. He founded the magazine *Science & Spirit*. His current research focuses on both developing an adequate spiritual understanding for our modern world and on pioneering studies of Paleolithic finger lines found in caves.

Rebecca Bryant holds a Ph.D. in philosophy from the University of Edinburgh. She is author of a previous book, *Discovery and Decision: Exploring the Metaphysics and Epistemology of Scientific Classification* (2000), plus numerous articles in the fields of philosophy and science and religion. She is a design and production manager at Oxford University Press.